Supplementary Exercises

for

Principles of Chemistry

Student's Version

Supplementary Exercises

for

Principles of Chemistry

Student's Version

MICHAEL MUNOWITZ

W • W • NORTON & COMPANY   NEW YORK • LONDON

ISBN 0-393-97549-5

W. W. Norton & Company, Inc., 500 Fifth Avenue, New York, N.Y. 10110
http://www.wwnorton.com

W. W. Norton & Company Ltd., 10 Coptic Street, London WC1A 1 PU

2 3 4 5 6 7 8 9 0

Contents

Chapter 1

Fundamental Concepts

S1-1. Light travels at a speed of 3.00×10^8 m s^{-1} in vacuum. Calculate the equivalent value in each of the following sets of units:

(a) kilometers per minute
(b) centimeters per week
(c) miles per hour
(d) inches per year

S1-2. The dimensions of a certain cube are 5.00 cm × 5.00 cm × 5.00 cm. Calculate the volume in liters.

S1-3. Calculate the volume, in milliliters, of a sphere that has a radius of 1.00 m.

S1-4. The *acre*, a unit of area, is equal to 4840 square yards. **(a)** Calculate the equivalent value in square feet. **(b)** Calculate the equivalent value in square meters.

S1-5. Confined at atmospheric pressure and a temperature of 0°C, a gas containing 6.0×10^{23} helium atoms will fill a cubical container 28.2 cm on a side. How many atoms are distributed per milliliter?

S1-6. Assume that a body has a net positive charge of 1.00 C. How many electrons must be added to make the body neutral?

S1-7. Two stationary particles, each with a charge of –2.00 C, interact at a distance of 1.00 cm. **(a)** Calculate the electrostatic force. **(b)** At what distance will the electrostatic force fall to 6.25% of its original value? **(c)** At what distance will the electrostatic force be tripled relative to its original value?

1

S1-8. A 5.00-kg body is raised 3.00 m in the earth's gravitational field. **(a)** How much work is done? **(b)** How much work would be done if the distance were doubled?

S1-9. Suppose that a perfectly round, homogeneous rubber ball falls freely along a line perpendicular to the earth's surface, its initial trajectory directed toward the center of the planet. Are there any circumstances under which the ball will bounce away at a 45° angle? Explain.

S1-10. Why do some balls bounce higher than others when released from the same height?

S1-11. Why does a ball bounce higher if it is *thrown* (rather than simply dropped) to the ground?

S1-12. Both a feather and a truck are released from a height of 100 m above the earth. Which object hits the ground first?

S1-13. The same feather and truck are released from a height of 100 m above the moon. **(a)** Which object hits the lunar surface first? **(b)** Which object delivers more momentum when it hits? **(c)** Which object is moving faster when it hits?

S1-14. A 1000-kg mass is dropped from a height of 10 m above the earth and also from a height of 10 m above the moon. Which surface—terrestrial or lunar—receives a larger dose of momentum?

S1-15. For each combination of units, state whether the quantity can possibly represent either velocity, acceleration, momentum, or energy:

$$\textbf{(a)}\ \text{lb g}^{-1} \qquad \textbf{(b)}\ \text{lb mi h ft}^{-1} \qquad \textbf{(c)}\ \text{lb mi}^3\,\text{ft}^{-1}\,\text{h}^{-1} \qquad \textbf{(d)}\ \text{lb min kg}^{-1}$$

Note that the *pound* (lb) is a unit of force in the English system.

S1-16. Show that the statement below is dimensionally correct:

$$\text{Energy} \sim \frac{(\text{momentum})^2}{\text{mass}}$$

S1-17. What quantity is expressed in units of *foot-pounds* (ft lb)?

S1-18. Show that the statement below is dimensionally correct:

$$\text{Momentum} \sim \frac{\text{energy} \times \text{time}}{\text{length}}$$

S1-19. The *watt*, used to measure electric power, represents the quantity *work per unit time*. Express the watt in terms of kg, m, and s.

S1-20. The electric field, defined as force per unit charge, is expressed naturally as *newtons per coulomb*. Show that the electric field can also be stated in units of *volts per meter*.

S1-21. The acceleration due to gravity (g) is approximately 9.81 m s^{-2} near the surface of the earth, a body with a mass of 5.98×10^{27} g. Estimate the radius of the earth, assuming the planet to be a sphere. Note that the value of the universal gravitational constant (G) is 6.67×10^{-11} m^3 kg^{-1} s^{-2}.

S1-22. The moon has a radius of 1.74×10^6 m, and the acceleration due to lunar gravity is 1.62 m s^{-2}. Estimate the mass of the moon.

S1-23. Suppose that Planet X has a mass of 1.00×10^{26} kg and a radius of 1.00×10^8 m. Estimate the acceleration due to gravity near the surface of Planet X.

Chapter 2

Atoms and Molecules

S2-1. How many protons, neutrons, and electrons are contained in each of the following isotopes?

$$\text{(a) } ^{33}\text{S} \qquad \text{(b) } ^{34}\text{S} \qquad \text{(c) } ^{36}\text{S}$$

S2-2. How many protons, neutrons, and electrons are contained in each of the following isotopes?

$$\text{(a) } ^{28}\text{Si} \qquad \text{(b) } ^{29}\text{Si} \qquad \text{(c) } ^{30}\text{Si}$$

S2-3. A certain element X has four isotopes: ^{54}X, ^{56}X, ^{57}X, ^{58}X. The ion X^{3+} contains a total of 23 electrons. **(a)** Identify X. **(b)** State the number of protons and neutrons in each isotope.

S2-4. Chromium exists as four isotopes: ^{50}Cr, ^{52}Cr, ^{53}Cr, ^{54}Cr. Use the information in the table below to determine the molar mass of chromium in its natural state.

	MOLAR MASS (g mol^{-1})	ABUNDANCE (%)
^{50}Cr	49.946046	4.345
^{52}Cr	51.940509	83.789
^{53}Cr	52.940651	9.501
^{54}Cr	53.938882	2.365

5

S2-5. The element boron exists as two isotopes: ^{10}B, ^{11}B. Use the information in the table below to determine the molar mass of ^{10}B.

	MOLAR MASS (g mol^{-1})	ABUNDANCE (%)
^{10}B	—	19.9
^{11}B	11.009305	80.1

Consult the periodic table as needed.

S2-6. How many core electrons and valence electrons are contained in each of the following atoms and ions?

(a) Ca   **(b)** Ti^{2+} **(c)** Ba   **(d)** Fr   **(e)** Al

S2-7. How many core electrons and valence electrons are contained in each of the following atoms?

(a) N   **(b)** P   **(c)** O   **(d)** S   **(e)** F   **(f)** Cl

S2-8. How many core electrons and valence electrons are contained in each of the following molecules?

(a) N$_2$ **(b)** O$_2$ **(c)** F$_2$ **(d)** Cl$_2$

S2-9. Which of the following atoms are likely to form anions? Which are likely to form cations?

(a) Cs   **(b)** O   **(c)** Cr   **(d)** I   **(e)** Ra

S2-10. A certain molecule consists of one atom of carbon and four atoms of iodine. **(a)** Draw the Lewis structure. **(b)** Use the VSEPR model to predict the most likely geometry.

S2-11. A certain molecule consists of one atom of beryllium and two atoms of chlorine. **(a)** Draw the Lewis structure. **(b)** Use the VSEPR model to predict the most likely geometry.

S2-12. **(a)** Draw a Lewis structure for the molecule GeF$_2$. Assume that germanium contributes four valence electrons. **(b)** According to the VSEPR model, how many valence electrons are present on the central atom? How are they arranged? **(c)** What is the expected molecular geometry of GeF$_2$?

S2-13. **(a)** Draw a Lewis structure for the molecule ClF_3. **(b)** According to the VSEPR model, how many pairs of valence electrons are present on the central atom? How are they arranged? **(c)** What is the expected molecular geometry?

S2-14. **(a)** Draw a Lewis structure for the molecular ion ICl_2^-. Assume that the iodine atom contributes seven valence electrons. **(b)** According to the VSEPR model, how many pairs of valence electrons are present on the central atom? How are they arranged? **(c)** What is the expected molecular geometry?

S2-15. **(a)** Draw a Lewis structure for the molecular ion $SnCl_3^-$. Assume that tin contributes four valence electrons. **(b)** According to the VSEPR model, how many pairs of valence electrons are present on the central atom? How are they arranged? **(c)** What is the expected molecular geometry?

S2-16. If the price of gold is \$300 per troy ounce (31.1 g), what is the value of a single atom?

S2-17. If the price of silver, expressed in British pounds sterling, is £3.00 per troy ounce (31.1 g), what is the equivalent price in dollars per mole? Assume an exchange rate of £1.00 = \$1.60.

S2-18. How many atoms of copper are there in a metric ton of the metal (1000 kg)?

S2-19. The Milky Way contains on the order of 100 billion stars. **(a)** Approximately how many "moles of stars" are present in our galaxy? **(b)** If each star had the mass of a hydrogen atom, what would be the total mass of the Milky Way?

S2-20. A Robber Baron exploits the masses at the rate of 1 million dollars per hour. Approximately how many years will the Robber Baron need to accumulate a fortune equal to 1 mole of dollars?

S2-21. Calculate the percent composition of each element in the following compounds:

  (a) $C_6H_{12}O_6$ **(b)** Al_2O_3 **(c)** Ag_2SO_4 **(d)** Na_2Te

Which of the compounds are molecular and which are ionic?

S2-22. Calculate the percent composition of each element in the following compounds:

  (a) N_2O_5 **(b)** C_6H_5OH **(c)** CH_3COOH **(d)** $ZnGa_2O_4$

Which of the compounds are molecular and which are ionic?

S2-23. Use the elemental analysis given below to determine an empirical formula:

  Na: 18.78%   Cl: 28.95%   O: 52.27%

S2-24. The molar mass of an unknown molecular compound is 56.11 g mol^{-1}. **(a)** Use the elemental data tabulated below to determine an empirical formula:

C: 85.63%   H: 14.37%

(b) What is the molecular formula?

S2-25. Balance the following equation:

$$\underline{} HNO_3 \rightarrow \underline{} H_2 + \underline{} N_2 + \underline{} O_2$$

S2-26. Balance the following equation:

$$\underline{} NH_3 + \underline{} O_2 \rightarrow \underline{} NO + \underline{} H_2O$$

S2-27. Consider the reaction below:

$$BF_3 + NH_3 \rightarrow BF_3NH_3$$

(a) How many grams of BF_3 will react completely with 50.00 g NH_3? **(b)** How many grams of BF_3NH_3 will be produced as a result?

S2-28. The gas-phase reaction of ozone and nitric oxide is described by the following chemical equation:

$$O_3 + NO \rightarrow NO_2 + O_2$$

(a) Suppose that equal masses of O_3 and NO are present. Which of the two reactants is stoichiometrically limiting? **(b)** How many grams of NO_2 and O_2 are produced when 10.00 g NO are completely consumed?

S2-29. Nitric oxide reacts with molecular bromine according to the following chemical equation:

$$2NO + Br_2 \rightarrow 2NOBr$$

(a) How many grams of NOBr are produced when 50.00 g NO react with 50.00 g Br_2? **(b)** Which of the two reactants is consumed completely? **(c)** How many grams of the other reactant remain unconsumed?

Chapter 3

Prototypical Reactions

S3-1. Identify the Brønsted-Lowry acid and base in each pair:

(a) OH^-, H_2O (b) H_2SO_4, HSO_4^- (c) NH_3, NH_4^+ (d) O^{2-}, OH^-

S3-2. Write a neutralization reaction in aqueous solution from which each of the following salts may arise:

(a) K_2SO_4 (b) $LiNO_3$ (c) $NaBr$ (d) $BaCl_2$

S3-3. Which statement best describes the properties of water?

(a) H_2O is a Brønsted-Lowry acid.

(b) H_2O is a Brønsted-Lowry base.

(c) H_2O is neither a Brønsted-Lowry acid nor a Brønsted-Lowry base.

(d) H_2O is both a Brønsted-Lowry acid and a Brønsted-Lowry base.

Write the appropriate chemical equations to justify your choice.

S3-4. Which statement best describes the properties of the oxide ion, O^{2-}?

(a) O^{2-} is a Brønsted-Lowry acid.

(b) O^{2-} is a Brønsted-Lowry base.

(c) O^{2-} is neither a Brønsted-Lowry acid nor a Brønsted-Lowry base.

(d) O^{2-} is both a Brønsted-Lowry acid and a Brønsted-Lowry base.

Write the appropriate chemical equations to justify your choice.

S3-5. Which statement best describes the properties of the hydronium ion?

(a) H_3O^+ is a Brønsted-Lowry acid.

(b) H_3O^+ is a Brønsted-Lowry base.

(c) H_3O^+ is neither a Brønsted-Lowry acid nor a Brønsted-Lowry base.

(d) H_3O^+ is both a Brønsted-Lowry acid and a Brønsted-Lowry base.

Write the appropriate chemical equations to justify your choice.

S3-6. (a) Is the hydride anion, H^-, a Lewis acid or a Lewis base? Is it both? Is it neither?
(b) Is the hydrogen cation, H^+, a Lewis acid or a Lewis base? Is it both? Is it neither?

S3-7. (a) Is the chloride anion, Cl^-, a Lewis acid or a Lewis base? Is it both? Is it neither? **(b)** Is the sodium cation, Na^+, a Lewis acid or a Lewis base? Is it both? Is it neither?

S3-8. In which of the following reactions does redox take place?

(a) $N_2(g) + O_2(g) \rightarrow 2NO(g)$

(b) $Ag^+(aq) + Cl^-(aq) \rightarrow AgCl(s)$

(c) $2Ag(s) + Zn^{2+}(aq) + 2OH^-(aq) \rightarrow Ag_2O(s) + Zn(s) + H_2O(\ell)$

(d) $NH_3(aq) + H_2O(\ell) \rightarrow NH_4^+(aq) + OH^-(aq)$

S3-9. Determine the oxidation number of the transition-metal atom in each species:

  (a) $K_4[Fe(CN)_6]$ **(b)** $Na_2[MoOCl_4]$ **(c)** $[Pt(NH_3)_2BrCl]$ **(d)** $[CoF_6]^{3-}$

S3-10. Complete and balance the following equations:

(a) $F_2 \rightarrow F^-$

(b) $Cu^{2+} \rightarrow Cu$

(c) $Cu^{2+} \rightarrow Cu^{3+}$

(d) $OH + e^- \rightarrow$

Identify each reaction as an oxidation or reduction process. How many moles of electrons are lost or gained?

S3-11. How many moles of cations and anions will be produced when one mole of each of the following compounds is dissolved in water?

$$\textbf{(a) } Na_2SO_4 \qquad \textbf{(b) } Ce_2(SO_4)_3 \qquad \textbf{(c) } CsCN \qquad \textbf{(d) } LiI$$

Identify the cation and anion in each case.

S3-12. Each of the following formulas is incorrect:

$$\textbf{(a) } Na_2Cl \qquad \textbf{(b) } NaK \qquad \textbf{(c) } NaBr_2 \qquad \textbf{(d) } LiSO_4$$

Explain why.

S3-13. What mass of $LiNO_3$ is needed to produce 375 mL of a 1.07 M aqueous solution?

S3-14. A volume of 32.8 mL is drawn from a 1.76 M solution of Na_2SO_4. How many individual sodium cations and sulfate anions are contained in the 32.8 mL?

S3-15. Into what total volume of water must 4.36 g KBr be dissolved to produce a 0.578 M solution?

S3-16. A certain volume of 1.00 M $AgNO_3$ is mixed with 50.0 mL of 1.00 M NaCl, resulting in the precipitation of silver chloride. How many milliliters of silver nitrate must be added to produce 1.00 g AgCl? Note that NaCl is present in excess.

S3-17. The volume of a 3.00 M solution is doubled. What happens to the concentration?

S3-18. A 2.00 M solution loses one-quarter of its solvent owing to evaporation. What happens to the concentration?

S3-19. Suppose that you have 100 mL of a 0.100 M solution of K_2SO_4, from which you must prepare 10 mL of 0.0500 M K_2SO_4. How would you do so?

S3-20. Identify the free radical (if any) in each pair:

$$\textbf{(a) } N, N_2 \qquad \textbf{(b) } Ar, F \qquad \textbf{(c) } H, H^+ \qquad \textbf{(d) } H, H^-$$

S3-21. Identify the free radical (if any) in each pair:

$$\textbf{(a) } Cl, Cl^- \qquad \textbf{(b) } Na, Na^+ \qquad \textbf{(c) } O^-, O^{2-} \qquad \textbf{(d) } He, Li^+$$

S3-22. Complete and balance each of the following reactions:

(a) $CCl_2F_2 \rightarrow CClF_2 + \underline{}$

(b) $Cl + \underline{} \rightarrow ClO$

(c) $I_2 \rightarrow I$

Which reactants and products are free radicals? Draw Lewis structures for each.

Chapter 4

Light and Matter—Waves and/or Particles

S4-1. The speed of light in vacuum is 670.6 million miles per hour. Calculate the frequency (in s^{-1}) corresponding to each of the following electromagnetic wavelengths:

$\qquad$ **(a)** 1.00 mi $\qquad$ **(b)** 1.00 ft $\qquad$ **(c)** 1.00 in

S4-2. One *light-year* is defined as the distance traversed in one year by electromagnetic radiation traveling in vacuum. What is its value in **(a)** meters, **(b)** kilometers, **(c)** centimeters, **(d)** inches, **(e)** feet, and **(f)** miles?

S4-3. **(a)** What distance does a beam of light traveling in vacuum advance in 1.000 s? **(b)** Suppose that the beam bounces back and forth between two mirrors separated by 0.500 m. How much time is required for one round trip?

S4-4. Choose an arbitrary wavelength and frequency for a traveling wave. **(a)** Sketch three cycles of the oscillation, indicating the extent of the wavelength on the diagram. **(b)** Sketch three cycles for a wave traveling at the same speed but half the original frequency. **(c)** Sketch three cycles for a wave traveling at the same speed but twice the original frequency.

S4-5. Suppose that the frequency of one electromagnetic wave is ν_1 and the frequency of another electromagnetic wave traveling at the same speed is ν_2. What is the ratio of their wavelengths?

S4-6. Say that two pulses of red light originate from two different sources but have identical wavelengths ($\lambda = 700$ nm). They meet at the same point and at the same time, with one pulse having traveled a distance L_1 and the other having traveled a distance L_2. **(a)** If the signals arrive exactly 180° out of phase, what is the minimum value of the difference $|L_2 - L_1|$? **(b)** Repeat the calculation for a phase difference of 90°. **(c)** Do it again for 270°.

S4-7. A source produces X rays with a frequency of 3.00×10^{18} Hz. **(a)** What is the period of the wave (the time per cycle)? **(b)** How much of a time lag will produce a phase difference of 30° between two such waves?

S4-8. (a) Use the equation $E = h\nu$ to express energy in units of kg, m, and s. **(b)** Use the equation $p = h/\lambda$ to express momentum in units of kg, m, and s. **(c)** Show that the combination pc has units of energy. **(d)** Show that the combination mc has units of momentum.

S4-9. Is it possible to have an equation in which a quantity with dimensions *energy × time* is set equal to a quantity with dimensions *momentum × length*?

S4-10. Define a quantity $\bar{\nu}$ called the *wavenumber*:

$$\bar{\nu} = \frac{1}{\lambda}$$

(a) Show that the wavenumber is directly proportional to photon energy:

$$E = hc\bar{\nu}$$

(b) Compute the value of $\bar{\nu}$, in units of cm^{-1}, for a photon with frequency equal to 3.00×10^{14} Hz.

S4-11. (a) Compute the energy of a photon with a wavenumber of 1000 cm^{-1}. **(b)** Do the same for a wavenumber of 500 cm^{-1}. **(c)** Which of the two photons has the higher frequency? **(d)** Which of the two photons has the longer wavelength?

S4-12. Which photons carry more energy?

(a) X ray or gamma ray
(b) red or green
(c) microwave or infrared
(d) violet or ultraviolet

S4-13. Which photons carry more momentum?

(a) blue or yellow
(b) infrared or ultraviolet
(c) radio or microwave
(d) red or infrared

S4-14. (a) A pulse of light delivers 1.00 J of electromagnetic energy at a wavelength of 1987 Å. Approximately how many photons are contained in the pulse? **(b)** Approximately how many photons are contained in a pulse of the same frequency that delivers only 1.00×10^{-16} J?

S4-15. Chromium has a work function of 7.21×10^{-19} J. Calculate the maximum wavelength able to expel a photoelectron.

S4-16. Illuminated at a wavelength of 4672 Å, a surface of europium emits a photoelectron with kinetic energy equal to 2.47×10^{-20} J. Calculate the work function of europium.

S4-17. The work function of cobalt is 8.01×10^{-19} J. Calculate the electromagnetic wavelength that will impart a velocity of 2.04×10^5 m s^{-1} to an outgoing photoelectron.

S4-18. Consider a one-dimensional standing wave confined over a distance of 1 meter. **(a)** Sketch the third harmonic. **(b)** How many nodes are present? Specify, in meters, the location of each node.

S4-19. Rank the following particles in order of increasing de Broglie wavelength, assuming each to be moving at the same velocity: proton, electron, neutron, deuterium nucleus, methane molecule (CH_4).

S4-20. Assume that an electron is moving at 1.00×10^6 m s^{-1}. At what velocity will a neutron have the same de Broglie wavelength? Note that the mass of a neutron is 1.675×10^{-27} kg.

S4-21. An alternative form of the indeterminacy principle expresses the relationship between uncertainty in energy and uncertainty in time:

$$\Delta E \, \Delta t \geq \frac{h}{4\pi}$$

Suppose that a particular quantum state has a lifetime of 1.0×10^{-8} s. Estimate the corresponding uncertainty in its energy.

Chapter 5

Quantum Theory of the Hydrogen Atom

S5-1. Pick the subshell in which a hydrogen electron will have the larger orbital angular momentum:

(a) 2p, 4s **(b)** 3p, 3d **(c)** 1s, 2p **(d)** 4f, 8s

S5-2. Repeat the exercise for He$^+$. In which subshell will the electron have the larger orbital angular momentum?

(a) 2p, 4s **(b)** 3p, 3d **(c)** 1s, 2p **(d)** 4f, 8s

S5-3. Pick the subshell in which a hydrogen electron will have the larger energy:

(a) 4d, 5s **(b)** 2s, 2p **(c)** 3d, 6s **(d)** 6s, 6p

S5-4. Again, for He$^+$: In which subshell will the electron have the larger energy?

(a) 4d, 5s **(b)** 2s, 2p **(c)** 3d, 6s **(d)** 6s, 6p

S5-5. How many spatial orbitals are contained within the $n = 7$ shell of a hydrogen atom? List them.

S5-6. How many values of m_ℓ correspond to the quantum number $\ell = 5$? List them.

S5-7. How many angular nodes are present in a 7g orbital? How many radial nodes?

S5-8. State the values of n and ℓ that correspond to each subshell:

(a) 5g **(b)** 7s **(c)** 6p **(d)** 9d

S5-9. For each subshell in the preceding exercise, list all possible values of m_ℓ:

(a) $5g$ **(b)** $7s$ **(c)** $6p$ **(d)** $9d$

S5-10. Explain why each of the following combinations of n and ℓ is forbidden:

(a) $2f$ **(b)** $4g$ **(c)** $3h$ **(d)** $1p$

S5-11. Give the name of the subshell corresponding to each combination of n and ℓ:

(a) $n = 7$, $\ell = 5$
(b) $n = 4$, $\ell = 2$
(c) $n = 6$, $\ell = 0$
(d) $n = 5$, $\ell = 1$

S5-12. Which sets of quantum numbers are forbidden?

(a) $n = -1$, $\ell = 0$, $m_\ell = 0$, $m_s = \frac{1}{2}$
(b) $n = 1$, $\ell = 1$, $m_\ell = 0$, $m_s = \frac{1}{2}$
(c) $n = 5$, $\ell = 4$, $m_\ell = 5$, $m_s = -\frac{1}{2}$
(d) $n = 5$, $\ell = 5$, $m_\ell = 4$, $m_s = -\frac{1}{2}$

For each violation, state the reasons why.

S5-13. Which sets of quantum numbers are forbidden?

(a) $n = 1$, $\ell = 0$, $m_\ell = 0$, $m_s = \frac{1}{4}$
(b) $n = 0$, $\ell = 1$, $m_\ell = 0$, $m_s = \frac{1}{2}$
(c) $n = 3$, $\ell = 2$, $m_\ell = -3$, $m_s = -\frac{1}{2}$
(d) $n = \frac{3}{2}$, $\ell = \frac{1}{2}$, $m_\ell = \frac{1}{2}$, $m_s = \frac{1}{2}$

For each violation, state the reasons why.

S5-14. Which sets of quantum numbers are forbidden?

(a) $n = 1$, $\ell = 0$, $m_\ell = 0$, $m_s = 1$
(b) $n = 1$, $\ell = 1$, $m_\ell = 0$, $m_s = 0$
(c) $n = 3$, $\ell = 2$, $m_\ell = -2$, $m_s = -\frac{1}{2}$
(d) $n = 10{,}000$, $\ell = 1$, $m_\ell = 0$, $m_s = \frac{1}{2}$

For each violation, state the reasons why.

S5-15. The *Bohr theory*, an early model of the one-electron atom, treats the electron as a "planet" orbiting a nuclear "sun" in a quantized, electromagnetic solar system. By restricting the electronic angular momentum to the values $nh/2\pi$ (where $n = 1, 2, 3, \ldots$), Bohr argued that the electron can persist indefinitely in circular orbits of fixed radii—in clear defiance of the laws of classical mechanics and electromagnetism. The idea worked, too, but only up to a certain point: Although the quantized energies he predicted were correct, this "old quantum theory" of fixed orbits soon proved unsuitable for systems more complex than one-electron atoms. In what way is the Bohr model in conflict with our present-day understanding of quantum mechanics?

S5-16. The radius of a Bohr orbit is given by the formula below:

$$r_n = \frac{\varepsilon_0 n^2 h^2}{\pi Z e^2 m_e} = \frac{n^2}{Z} a_0$$

The symbols ε_0, h, e, and m_e denote, respectively, the permittivity of vacuum (Chapter 1), Planck's constant, the magnitude of the electronic charge, and the mass of the electron. Numerical values for all these physical constants are tabulated in Appendix C of *Principles of Chemistry*. **(a)** Compute a_0. **(b)** Compute the orbital radius of a hydrogen electron in the first shell. **(c)** Do the same for a hydrogen electron in the second shell. **(d)** Once more: for a hydrogen electron in the third shell.

S5-17. Repeat the preceding exercise for He$^+$: Calculate r_1, r_2, and r_3, and explain the trend.

S5-18. Do the same for Li^{2+}.

S5-19. A Bohr electron orbiting with radius r_n has the following velocity:

$$v_n = \frac{Ze^2}{2\varepsilon_0 nh}$$

Write an equation for the corresponding kinetic energy.

S5-20. The electrostatic potential energy between the electron and nucleus in a one-electron atom is given below:

$$E_{pot} = -\frac{Ze^2}{4\pi\varepsilon_0 r}$$

(a) Combine this expression with the kinetic energy of an orbiting Bohr electron (from the preceding exercise) to show that the total energy is quantized as follows:

$$E_n = -\frac{m_e Z^2 e^4}{8\varepsilon_0^2 n^2 h^2}$$

(b) Insert values of the physical constants into the expression in (a), and thus show that E_n is numerically equivalent to the consolidated formula below:

$$E_n = -\frac{Z^2}{n^2} R_\infty$$

S5-21. Consider again the energy equation,

$$E_n = -\frac{m_e Z^2 e^4}{8\varepsilon_0^2 n^2 h^2}$$

and note that the Balmer wavelengths for hydrogen are 656 nm, 486 nm, 434 nm, and 410 nm. What emission wavelengths would be observed if the mass of an electron were suddenly doubled?

S5-22. Repeat the preceding exercise, but this time pretend that the *charge* of an electron is doubled from its usual value. In what ways will the Balmer wavelengths for hydrogen be affected?

S5-23. Imagine, once more, what would happen if the physical constants m_e and e were to depart from their usual values. **(a)** Compute the ionization energies for H and He^+ that would result if the mass of the electron were doubled. Explain the relationship. **(b)** Compute the ionization energies that would result if the electronic charge were doubled.

Chapter 6

Periodic Properties of the Elements

S6-1. Count the total number of s electrons in each of the following atoms (inclusive of core and valence):

(a) Ta **(b)** Ag **(c)** Cs **(d)** K **(e)** In

S6-2. Count the total number of p electrons in each of the following atoms (inclusive of core and valence):

(a) Ar **(b)** I **(c)** W **(d)** Ra **(e)** Li

S6-3. Count the total number of d electrons in each of the following atoms (inclusive of core and valence):

(a) P **(b)** Ti **(c)** Sn **(d)** La **(e)** Co

S6-4. Count the total number of f electrons in each of the following atoms (inclusive of core and valence):

(a) La **(b)** Lu **(c)** Pt **(d)** Rf **(e)** Xe

S6-5. Name an atom that contains a partially filled f subshell.

S6-6. Name an element from the d block.

S6-7. Name an element that contains a filled valence p subshell.

S6-8. Name two elements that have an s^1 valence configuration.

S6-9. **(a)** Write the electron configuration for a hydride ion, H^-. **(b)** Write the electron configuration for a helium atom, He. **(c)** Which species is more reactive, H^- or He? Explain why.

S6-10. Write electron configurations for the following atoms:

> **(a)** Tc   **(b)** I   **(c)** Ag   **(d)** Rb   **(e)** Zn

S6-11. Write electron configurations for the following atoms:

> **(a)** Xe   **(b)** Cs   **(c)** Ca   **(d)** Tl   **(e)** Bi

S6-12. Identify the atom from the valence configuration stated:

> **(a)** $7s^1$ **(b)** $5s^2$ **(c)** $4s^2 3d^{10} 4p^3$ **(d)** $1s^2$ **(e)** $2s^2 2p^1$

S6-13. Identify the atom from the valence configuration stated:

> **(a)** $4s^2 3d^3$ **(b)** $3s^2 3p^6$ **(c)** $6s^1$ **(d)** $6s^2$ **(e)** $6s^2 5d^1$

S6-14. Which of the following atoms are paramagnetic?

> **(a)** He   **(b)** Cs   **(c)** Sr   **(d)** Br   **(e)** Ga

S6-15. Count the number of unpaired electrons in each atom:

> **(a)** Ge   **(b)** Kr   **(c)** Fr   **(d)** P   **(e)** Ti

S6-16. Which species are diamagnetic?

> **(a)** H^- **(b)** H   **(c)** Cl^- **(d)** Cl   **(e)** Ar

S6-17. Which species are diamagnetic?

> **(a)** Mg^+ **(b)** Ca^{2+} **(c)** Li^+ **(d)** Al^+ **(e)** Sc

S6-18. Show what the periodic table would look like if the transition metals were separated from its main body—like the lanthanide and actinide elements.

S6-19. Show what the periodic table would look like if the lanthanide and actinide series were incorporated into its main body—like the transition metals.

S6-20. Write the electron configuration of element 119, as yet undiscovered. Where would it fall in the periodic table?

S6-21. To which period of the periodic table does the element europium belong—fourth, fifth, sixth, or seventh?

S6-22. To which period of the periodic table does the element americium belong—fourth, fifth, sixth, or seventh?

S6-23. How many elements would an eighth row of the periodic table contain?

S6-24. Arrange the following atoms in order of increasing radius:

(a) Sr, F, Li, Rb
(b) Al, Cl, H, Cs
(c) Ca, K, Fr, Si
(d) Se, F, Rn, Xe

S6-25. Arrange the following ions in order of increasing radius:

(a) Br^-, Li^+, Be^+, Be^{2+}
(b) Na^+, Mg^{2+}, Al^{3+}, Cl^-
(c) Br^-, I^-, O^{2-}, H^-
(d) Ca^{2+}, Sr^{2+}, Cs^+, Ba^{2+}

S6-26. Pick an atom, perhaps one of many, that fits each of the following descriptions:

(a) radius smaller than Br
(b) radius greater than Rn
(c) radius greater than K
(d) radius smaller than O

S6-27. Pick an atom, perhaps one of many, that fits each of the following descriptions:

(a) ionization energy greater than H
(b) ionization energy less than K
(c) ionization energy greater than P
(d) ionization energy less than Na

S6-28. Pick the species with the larger ionization energy:

(a) Li, He   **(b)** Li, Li^+ **(c)** Li^+, He   **(d)** Li^{2+}, He^+

S6-29. Arrange the following atoms in order of increasing ionization energy:

(a) Mg, Sr, He   **(b)** F, Ba, Cl   **(c)** Br, Rb, Li   **(d)** Ne, Si, P

S6-30. Pick the atom with the more favorable electron affinity:

(a) Cs, Mg   (b) Cs, Kr   (c) Sb, I   (d) I, Sr

S6-31. Explain why the alkaline earth metals have unfavorable electron affinities.

S6-32. Arrange the following elements (undiscovered as of the year 2000) in their expected order of (a) increasing ionization energy and (b) increasing atomic radius: 115, 117, 119.

S6-33. Why is helium assigned to Group VIII and not Group II?

S6-34. Noble gases are said to be inert, except in rare cases. If so, how do you account for the prevalence of neon signs? Isn't neon a noble gas?

Chapter 7

Covalent Bonding and Molecular Orbitals

S7-1. Suppose that a homonuclear diatomic system (second row) has an odd number of electrons. **(a)** Do you have sufficient information to determine whether the structure is a molecule or an ion? **(b)** If you were to determine that the structure is ionic, would you be able to tell whether it is an anion or a cation?

S7-2. Suppose that a heteronuclear diatomic system (second row) has an odd number of electrons. **(a)** Do you have sufficient information to determine whether the structure is a molecule or an ion? **(b)** If you were to determine that the structure is ionic, would you be able to tell whether it is an anion or a cation?

S7-3. Identify the homonuclear diatomic molecule (second row) that best fits each of the following descriptions:

(a) 6 bonding and 4 antibonding electrons (inclusive of core and valence)
(b) 4 bonding and 2 antibonding electrons (inclusive of core and valence)
(c) 2 bonding electrons (valence)
(d) 4 bonding and 2 antibonding electrons (valence)

S7-4. A homonuclear diatomic species (second row) has a bond order of 2.5. **(a)** Is the species a molecule or an ion? **(b)** Write a possible valence electron configuration. **(c)** Is your proposed species paramagnetic? If so, how many electrons are unpaired?

S7-5. Identify the homonuclear diatomic molecule (second row) that best fits each of the following descriptions:

(a) 4 bonding electrons (valence); bond order = 1; paramagnetic
(b) 6 antibonding electrons (valence); bond order = 1; diamagnetic
(c) 2 bonding electrons (valence); bond order = 1; diamagnetic

S7-6. Identify the homonuclear diatomic molecule (second row) that best fits each of the following descriptions:

(a) bond order = 2; paramagnetic
(b) bond order = 2; diamagnetic
(c) bond order = 3; diamagnetic

One of the two pieces of information given in part (c) is redundant. Which is it?

S7-7. Consider the following diatomic systems: Li_2^+, Li_2, Li_2^-.

(a) Calculate the bond order for each structure.
(b) Which structure has the largest dissociation energy?
(c) Which structure has the shortest bond length?

S7-8. Consider the following diatomic systems: Be_2^+, Be_2, Be_2^-.

(a) Calculate the bond order for each structure.
(b) Which structure has the smallest dissociation energy?
(c) Which structure has the longest bond length?

S7-9. Is it theoretically possible for a second-row, homonuclear diatomic system in its ground state to have a bond order equal to 4? If so, use an energy-level diagram to sketch a plausible electron configuration.

S7-10. Is it theoretically possible for a second-row, homonuclear diatomic system in an *excited* state to have a bond order equal to 4? If so, use an energy-level diagram to sketch a plausible electron configuration.

S7-11. **(a)** Explain the significance of bonding, antibonding, and nonbonding orbitals in the molecular orbital picture of chemical bonding. **(b)** Is the energy of a bonding molecular orbital greater than, less than, or equal to the energy of the two atomic orbitals from which it arises? **(c)** Same question, but for an antibonding orbital: Is the energy greater than, less than, or equal to the energy of the original atomic orbitals? **(d)** Once more, for a nonbonding molecular orbital: Is the energy greater than, less than, or equal to the energy of the atomic orbitals?

S7-12. Count the number of bonding, nonbonding, and antibonding electrons in the π system of each species:

(a) $CH_2CHCH_2^+$, the allyl cation
(b) CH_2CHCH_2, the allyl free radical
(c) $CH_2CHCH_2^-$, the allyl anion

S7-13. (a) Draw a Lewis structure for the molecule C_2Cl_2. **(b)** Identify each bond as σ or π. **(c)** Describe the bond angles and hybridization around each carbon. **(d)** Does the molecule have an overall dipole moment?

S7-14. Rank the following molecules in order of increasing dipole moment: C_2H_2, C_2Cl_2, C_2HF, C_2HCl.

S7-15. (a) Draw a Lewis structure for the molecule CH_3CH_2Br. **(b)** Identify each bond as σ or π. **(c)** Describe the bond angles and hybridization around each carbon.

S7-16. (a) Draw a Lewis structure for the molecule CH_2CHCH_3. **(b)** Identify each bond as σ or π. **(c)** Describe the bond angles and hybridization around each carbon.

S7-17. (a) Draw a Lewis structure for the molecule $CH_2CHCHCHCH_3$. **(b)** Identify each bond as σ or π. **(c)** Describe the bond angles and hybridization around each carbon.

S7-18. (a) Sketch the lowest delocalized π orbital for the molecule $CH_2CHCHCHCHCH_2$. **(b)** How many nodes does it contain? **(c)** Sketch the highest delocalized π orbital. **(d)** How many nodes does it contain?

S7-19. (a) Draw the Lewis structure of $BeCl_2$. **(b)** Use the VSEPR model to predict the shape of the molecule. **(c)** Is the dipole moment of the molecule greater than, less than, or the same as the dipole moment of BeF_2?

S7-20. (a) Use the VSEPR model to predict the geometry of the electron pairs in IF_5. How many bonding pairs and how many lone pairs are disposed about the central atom? **(b)** What is the likely shape of the molecule?

S7-21. (a) Draw three resonance structures for the nitrate ion, NO_3^-. **(b)** Use the VSEPR model to predict the shape of the ion. **(c)** Propose a hybridization scheme consistent with the geometry.

S7-22. The formate ion, HCO_2^-, is a planar structure with bond angles of approximately $120°$. **(a)** Draw two possible resonance forms. **(b)** Propose a hybridization scheme consistent with the geometry. **(c)** Describe the π bond. Is it localized or delocalized?

S7-23. Consider again the formate ion, HCO_2^-. **(a)** Sketch the lowest π molecular orbital. **(b)** Predict whether the bond order of each C—O linkage will fall between 0 and 1, between 1 and 2, or between 2 and 3. Take into account both σ and π electrons.

S7-24. (a) Use an energy-level diagram to sketch a possible π-electron configuration for the benzene anion, $C_6H_6^-$. **(b)** Is the ion more stable or less stable relative to neutral C_6H_6?

S7-25. Suppose that a molecule contains 10 atoms and 50 electrons. Is it theoretically possible for the electron density between two particular atoms to have the value 0.0647? What does it mean to speak of "647 ten-thousandths of an electron"?

S7-26. Fluorine is the most electronegative of all the elements. Are all heteronuclear bonds involving fluorine highly polar? Explain why or why not.

Chapter 8

Some Organic and Biochemical Species and Reactions

S8-1. Which structures may exist as open-chain alkanes?

    (a) $C_{17}H_{36}$ **(b)** $C_{25}H_{50}$ **(c)** C_8H_{16} **(d)** $C_{12}H_{22}$ **(e)** $C_{11}H_{24}$

S8-2. Which structures may exist as closed rings?

    (a) $C_{10}H_8$ **(b)** $C_{20}H_{42}$ **(c)** C_2H_2 **(d)** C_4H_8 **(e)** $C_6H_3Cl_3$

S8-3. Which structures may exist as planar molecules?

    (a) C_2H_2BrCl **(b)** $C_2H_4Br_2$ **(c)** $C_{10}H_8$ **(d)** C_7H_{16} **(e)** C_6H_{12}

S8-4. Which structures may contain multiple bonds?

    (a) $C_4H_8Br_2$ **(b)** $C_8H_{14}Cl_2$ **(c)** C_2Cl_6 **(d)** C_6H_6 **(e)** C_2H_6O

S8-5. Write a generic formula for an alkene containing n carbon atoms and m double bonds.

S8-6. Write a generic formula for an alkyne containing n carbon atoms and l triple bonds.

S8-7. Write a generic formula for a hydrocarbon containing n carbon atoms, m double bonds, and l triple bonds.

S8-8. Describe the structure and bonding of $CH_3CCCH_2CClCH_2$.

S8-9. Describe the structure and bonding of the aromatic molecule $C_6H_3Cl_3$.

S8-10. Draw any 10 structural isomers of decane, $C_{10}H_{22}$.

S8-11. How many structural isomers of $C_{18}H_{38}$ are derived from a chain of 17 carbon atoms?

S8-12. Three of the following four formulas represent exactly the same structure:

$$H_3C\!\!-\!\!CH_2\text{-}CH\!-\!CH_2\text{-}CH_2\text{-}CH_3 \qquad\qquad H_3C\!\!-\!\!CH_2\text{-}CH_2\text{-}CH\!-\!CH_2\text{-}CH_3$$
$$\overset{\displaystyle |}{CH_3} \qquad\qquad\qquad\qquad\qquad\qquad \overset{\displaystyle |}{CH_3}$$

$$\qquad\qquad\qquad\qquad\qquad\qquad\qquad \overset{\displaystyle CH_3}{\overset{\displaystyle |}{}}$$
$$H_3C\!\!-\!\!CH\!-\!CH_2\text{-}CH_2\text{-}CH_3 \qquad\qquad H_3C\!\!-\!\!\overset{\displaystyle |}{\underset{\displaystyle |}{C}}\!\!-\!\!CH_2\text{-}CH_2\text{-}CH_3$$
$$\overset{\displaystyle |}{CH_2} \qquad\qquad\qquad\qquad\qquad\qquad CH_3$$
$$\overset{\displaystyle |}{CH_3}$$

Which one is different?

S8-13. Draw the cis and trans isomers of 2-octene.

S8-14. Draw structures for the two enantiomers of cysteine, an amino acid.

S8-15. Consider bromobenzene, C_6H_5Br, a molecule formed when one hydrogen on a benzene ring is replaced by bromine. Is bromobenzene optically active?

S8-16. Draw the ortho, meta, and para isomers of dibromobenzene.

S8-17. Suppose that each of three hydrogens on a benzene ring is replaced by group X. Draw the isomers that may result.

S8-18. Suppose that each of two hydrogens on a benzene ring is replaced by group X and that one other hydrogen is replaced by group Y. Draw the isomers that may result.

S8-19. Consider another trisubstituted benzene: one hydrogen is replaced by group X; a second hydrogen is replaced by group Y; a third is replaced by group Z. Draw the isomers that may result.

S8-20. First, redraw each molecule to show the connections explicitly. Second, identify the functional group(s):

  (a) CH_3COOH **(b)** CH_3OCH_3 **(c)** CH_3CH_2Cl **(d)** $CH_3CH_2CH_2CHOHCH_3$

S8-21. First, redraw each molecule to show the connections explicitly. Second, identify the functional group(s):

    (a) $HCOOH$ **(b)** $CH_3CONHCH_3$ **(c)** CH_3COCH_3 **(d)** C_2H_4

S8-22. Write a reaction that includes $CH_3CH_2CH_2COOCH_2CH_2CH_3$ as a product.

S8-23. Write a reaction in which the reactants are CH_3COOH and CH_3CH_2COOH.

S8-24. Write a reaction that describes the condensation of methionine and alanine into a dipeptide.

S8-25. Write a reaction in which the reactants are $CH_3CH_2CH_2CHCHCH_2CH_3$ and H_2.

S8-26. S_N1 reactions are generally favored in strongly polar solvents. Suggest a reason why.

Chapter 9

States of Matter

S9-1. Temperature is a measure of the average kinetic energy of particles in motion. Would you expect the volume of a gas to increase, decrease, or remain the same as the temperature is increased? Assume that the gas remains under constant atmospheric pressure throughout.

S9-2. Would you expect the volume of a gas to increase, decrease, or remain the same as the system is subjected to increasing external pressure? Assume that the temperature remains constant.

S9-3. Under what conditions of temperature and pressure would you expect a gas to condense into a liquid?

S9-4. Under what conditions of pressure and volume would you expect a gas to condense into a liquid?

S9-5. **(a)** Estimate, to within a power of 10, the number of molecules contained in the air filling an average-sized bedroom. **(b)** Estimate the corresponding total mass of air. In each case, state the assumptions you make in order to arrive at the numbers you do. Turn to whatever sources of information you deem helpful.

S9-6. Which substance in each pair is more likely to exist as a gas at room temperature?

    (a) H_2O, C_2H_6 **(b)** Rn, Li   **(c)** C_3H_8, C_6H_{14} **(d)** NaCl, HCl

S9-7. What are the principal interactions holding together each of the following phases?

    (a) Mg(s)   **(b)** Cl_2(g)   **(c)** $MgCl_2$(s)   **(d)** $MgCl_2$(aq)

S9-8. What are the principal interactions holding together each of the following phases?

(a) $H_2(g)$ (b) $Cl_2(s)$ (c) $HCl(\ell)$ (d) $HCl(aq)$

S9-9. Describe the interactions responsible for maintaining the primary, secondary, and tertiary structures of a protein.

S9-10. Which substance in each pair is more likely to boil at the higher temperature?

(a) $CH_3CH_2CH_2CH_2CH_3$, $CH_3CH_2CH_2CH_3$

(b) $CH_3CH_2CH_2CH_2CH_2OH$, $CH_2OHCHOHCH_2CH_2CH_2OH$

(c) $C_2H_4Cl_2$, C_2H_6

(d) CH_3OCH_3, H_2O

S9-11. Which substance in each pair is more likely to boil at the higher temperature?

(a) Xe, Kr   (b) CH_3OH, CF_4 (c) C_3H_8, C_3F_8 (d) C_2F_6, C_2H_5F

S9-12. (a) Will a gas consisting of He atoms conduct electricity? (b) The term *plasma* refers to an electrically neutral gas consisting of electrons and ions. Will a helium plasma conduct electricity? (c) Under what conditions of temperature will a plasma form?

S9-13. (a) In what way is a plasma (see the preceding exercise) similar to a metal? How is it different? (b) In what way is a plasma similar to an electrolytic solution? In what way is it different?

S9-14. Which substance in each pair is more likely to have the higher lattice energy?

(a) $CaCl_2$, KCl   (b) NaCl, CsCl   (c) $SrCl_2$, $MgCl_2$

S9-15. Gold (Au) crystallizes in a cubic lattice with cell dimension equal to 4.079 Å and with density equal to 19.282 g cm^{-3}. Is the unit cell a face-centered, body-centered, or simple cubic structure?

S9-16. How many atoms of metallic silver are found in a spherical sample with radius equal to 0.107 m? Consult Appendix C for any necessary data.

S9-17. How many molecules of H_2O are found in a spherical sample with radius equal to 0.107 m? Consult Appendix C for any necessary data.

S9-18. Assume that 1.00 mol He occupies a volume of 22.4 L. (a) How many atoms of He are found in a spherical sample with radius equal to 0.107 m? (b) Compare the value obtained here with the values obtained in the preceding two exercises.

S9-19. How many grams of $MgCl_2$ must be dissolved in a total volume of 365 mL to produce a chloride concentration of 0.00763 M?

S9-20. A 50.0-mL solution of glucose ($C_6H_{12}O_6$) has a concentration of 0.436 M. To what volume must the solution be diluted in order to make the concentration 0.200 M?

S9-21. A mixture of gases contains 5.76 g He, 50.6 g N_2, and 84.4 g CO_2. Calculate the mole fraction of each component.

S9-22. A mixture of ethanol and methanol contains 1.76 g CH_3OH, equivalent to a mole fraction of 0.287. Calculate the mass of ethanol present.

S9-23. Suppose that the mole fraction of ethanol is 0.459 in a certain aqueous solution. Calculate the ratio—by mass—of ethanol to water in the solution.

Chapter 10

Macroscopic to Microscopic—Gases and Kinetic Theory

S10-1. Suppose that water is used to construct a barometer. Will a column of water that registers 1.0000 atm on a hot summer day be taller than, shorter than, or the same height as a column of water that registers 1.0000 atm on a freezing winter day? Consult the appropriate table in Appendix C for any necessary data.

S10-2. Suppose, instead, that ethanol is used to construct a barometer. Calculate the density of the liquid, given that a column of ethanol rises 42.9 ft when the ambient pressure is 1.00 atm at 20°C. Consult Appendix C for additional data, if necessary.

S10-3. Why do you think that mercury is a preferred material for barometers?

S10-4. Assume that an ideal gas is confined at STP. How many particles, on the average, are contained in spheres with the following radii?

    (a) 1.00 m   **(b)** 1.00 cm   **(c)** 1.00 mm   **(d)** 1.00 μm   **(e)** 10.0 nm

S10-5. Charles's law states that

$$\frac{V}{T} = \text{constant}$$

for fixed n and P. **(a)** What is the value of the constant in Charles's law for $n = 1.0000$ mol and $P = 1.0000$ atm? **(b)** What is its value for $n = 0.10000$ mol and $P = 10.000$ atm? **(c)** For $n = 10.000$ mol and $P = 1.0000$ atm? **(d)** Is the value constant for all choices of n and P? Why or why not?

S10-6. Boyle's law states that

$$PV = \text{constant}$$

for fixed n and T. **(a)** Under what conditions will the constant in Boyle's law have the magnitude R (the universal gas constant)? **(b)** Under what conditions will it have the value 1.0000 atm L? **(c)** Under what conditions will it have the value 1.0000 J?

S10-7. Calculate the ratio of the constants k_{Boyle} and $k_{Charles}$ in Boyle's law and Charles's law:

$$PV = k_{Boyle} \qquad (\text{fixed } n, T)$$

$$\frac{V}{T} = k_{Charles} \qquad (\text{fixed } n, P)$$

S10-8. The temperature of an ideal gas is increased from 200 K to 400 K. At the same time, the pressure is increased from 1.50 atm to 3.00 atm. What happens to the density—does it increase, decrease, or remain the same?

S10-9. The temperature of an ideal gas is decreased from 300°C to 150°C. At the same time, the pressure is decreased from 2.00 atm to 1.00 atm. What happens to the density—does it increase, decrease, or remain the same?

S10-10. A 50.0-g mass of gaseous O_2 fills a container at STP. **(a)** Calculate the volume of the container. **(b)** How many particles of Xe would be found in a container twice the volume?

S10-11. How many grams of gaseous CO_2 occupy a volume of 3.50 L at a pressure of 1.10 atm and a temperature of 25.0°C?

S10-12. A mixture of two ideal gases has a total pressure of 535.1 torr. If the mole fraction of one component is 0.327, what is the partial pressure of the other component?

S10-13. What additional information will you need to solve the following problem?

A mixture of two ideal gases has a total pressure of 535.1 torr. If the mole fraction of one component is 0.327, what is the density of the other component?

S10-14. What additional information will you need to solve the following problem?

A sample of ideal gas occupies a volume of 22.4 L at 1.00 atm. How many particles are present in the container?

S10-15. What additional information will you need to solve the following problem?

A certain ideal gas consists of diatomic molecules X_2. Identify X, given that the total mass of a sample at STP is 10.00 g.

S10-16. What additional information will you need to solve the following problem?

A 3.00-mol sample of an ideal gas has a pressure of 0.333 atm. Calculate the average translational kinetic energy per particle.

S10-17. The root-mean-square speed of He is 1531 m s^{-1} at a certain temperature. Calculate the temperature.

S10-18. The translational kinetic energy of H_2 is 4.00 kJ mol^{-1} at a certain temperature. Calculate the root-mean-square speed of He at the same temperature.

S10-19. A mixture of H_2, He, and N_2 occupies a volume of 11.2 L at STP. Calculate the average translational kinetic energy *per particle* that results for **(a)** H_2, **(b)** He, **(c)** N_2.

S10-20. A mixture of H_2, He, and N_2 occupies a volume of 44.8 L at STP. Calculate the root-mean-square speed that results for **(a)** H_2, **(b)** He, **(c)** N_2.

S10-21. A 20.0-L sample of neon gas, originally at STP, is compressed isothermally to a volume of 18.0 L. **(a)** Calculate the translational kinetic energy per mole after the compression. **(b)** Calculate the root-mean-square speed after the compression.

S10-22. The root-mean-square speed of a certain ideal gas falls by 50% in response to a halving of the volume. What is the corresponding change in pressure?

S10-23. Recall the definition of the *mean free path* for a gas particle: the average distance traveled between collisions. **(a)** Will the mean free path increase, decrease, or remain the same the same if P is doubled at constant T? **(b)** Will the mean free path increase, decrease, or remain the same if T is doubled at constant P? **(c)** Will the mean free path increase, decrease, or remain the same if n is halved at constant P and T?

S10-24. Consider, in anticipation of Chapter 11, an obvious flaw in our model of the ideal gas: The particles in a gas clearly *do* have intrinsic volume; atoms and molecules are not simply point masses. Assume, therefore, that each particle occupies a finite space but is otherwise unable to interact with any other particle. **(a)** Will the volume of this special kind of "real" gas be greater than, less than, or the same as the volume of an ideal gas? **(b)** Given that

$$P_{ideal}V_{ideal} = nRT$$

for an ideal gas, how would you expect the intrinsic particle volume to affect the pressure of a real gas (P_{real})? Will P_{real} be higher than, lower than, or the same as P_{ideal}? **(c)** Under what conditions of temperature and pressure will it be most reasonable to neglect the intrinsic volume of the gas particles?

S10-25. Consider another obvious flaw in our picture of an ideal gas: Atoms and molecules certainly do have the ability to interact. All the usual means of intermolecular interactions—hydrogen bonding, dipole–dipole, dispersion, and so forth—remain inherent in the internal structure of the particles. **(a)** How would you expect these intermolecular interactions to affect the pressure of a real gas? Will P_{real} be higher than, lower than, or the same as P_{ideal}? **(b)** Under what conditions of volume and temperature will it be most reasonable to neglect the interactions?

S10-26. For which gas in each pair do you expect the equation of state

$$PV = nRT$$

to be more accurate?

(a) He, Rn   **(b)** H_2O, CH_4 **(c)** H_2, CO_2

S10-27. Take into account the preceding three exercises, and pick the statement that best describes the relationship between P_{real} and P_{ideal}:

(a) P_{real} is always greater than P_{ideal}.
(b) P_{real} is always less than P_{ideal}.
(c) P_{real} is always equal to P_{ideal}.
(d) P_{real} may be greater than, less than, or equal to P_{ideal}.

Justify your answer.

S10-28. Explain why Dalton's law of partial pressures is valid for ideal gases.

S10-29. The ideal gas equation of state predicts that the volume of a gas shrinks to zero as the temperature approaches 0 K:

$$V = \frac{nRT}{P}$$

(a) Why is this prediction *always* wrong for a real gas? What happens instead? **(b)** For which of these two gases—carbon dioxide or helium—will the "ideal" prediction be more nearly correct at a given temperature? Why?

Chapter 11

Disorder–Order and Phase Transitions

S11-1. Is it possible for a *real* gas to remain a gas as the temperature approaches absolute zero, regardless of density?

S11-2. Suppose that 25.0 grams of gaseous methane (CH_4) fill a volume of 100.0 L at a pressure of 1.00 atm. Is it possible to force a condensation by reducing the volume below 100 L? Note that the critical temperature for methane is −82.62°C.

S11-3. The density and vapor pressure of diethyl ether ($C_4H_{10}O$) at 25°C are 0.7080 g mL^{-1} and 532.5 torr, respectively. Calculate the vapor pressure of a solution in which 0.500 gram of naphthalene ($C_{10}H_8$) is dissolved in 25.0 milliliters of diethyl ether.

S11-4. Calculate the total vapor pressure over a solution that contains equal volumes of benzene (C_6H_6) and toluene ($C_6H_5CH_3$) at 25°C, given the information below:

COMPONENT	DENSITY (g mL^{-1})	VAPOR PRESSURE (torr, 25°C)
benzene	0.8729	95.3
toluene	0.8647	28.5

S11-5. Calculate the total vapor pressure over a solution that contains equal masses of methanol (CH_3OH) and water at 25°C, given the information below:

COMPONENT	DENSITY (g mL^{-1})	VAPOR PRESSURE (torr, 25°C)
methanol	0.7872	127.5
water	0.99705	23.8

S11-6. The vapor pressure of X (a solvent) is lowered by 10.0% when 1.00 mole of Y (a molecular solute) is dissolved in 1.00 kilogram of X. Do you have sufficient data to determine the molar mass of Y? If not, what additional information is needed?

S11-7. An aqueous solution of NaBr freezes at a temperature of –1.15°C. Calculate the mass of dissolved NaBr per kilogram of solvent.

S11-8. An aqueous solution contains 5.00 g $Ca(NO_3)_2$ and boils at 374.15°C. Calculate the corresponding mass of solvent.

S11-9. An aqueous solution contains 15.00 g $Ce_2(SO_4)_3$ dissolved in 150. mL H_2O. Assume that the density of water is 1.00 g mL^{-1}. **(a)** Calculate the freezing point. **(b)** How many grams of $NaCH_3COO$ must be dissolved in 400. mL H_2O to produce the same freezing point?

S11-10. A particular aqueous solution of sucrose freezes at –0.75°C. At what temperature does it boil?

S11-11. Suppose that a certain aqueous solution freezes at –3.00°C. **(a)** Can you determine the corresponding boiling point without knowing whether the solute is molecular or ionic? **(b)** If so, do it: Calculate the boiling point.

S11-12. The density of acetone (C_3H_6O) is 0.7899 g mL^{-1} at 20°C. When 2.00 grams of naphthalene ($C_{10}H_8$) are dissolved in 20.0 milliliters of acetone, the boiling point rises from 56.0°C to 57.7°C. Calculate the molal boiling-point-elevation constant, K_b, for acetone.

S11-13. The freezing point of water falls by 0.412°C when 1.00 g of an unknown molecular solute is dissolved. Given the appropriate value of K_f, do you have sufficient data to determine the molar mass of the solute? If not, what additional information is needed?

S11-14. The boiling point of methanol rises by 0.628°C when 0.562 mol of a molecular solute is dissolved. Given the appropriate value of K_b, do you have sufficient data to determine the density of methanol? If not, what additional information is required?

S11-15. Make a rough sketch, from memory, of the phase diagram for a substance such as CO_2. Mark the following locations:

(a) solid, liquid, and gas regions
(b) solid–liquid, solid–gas, and gas–liquid coexistence regions
(c) critical isotherm
(d) supercritical region

S11-16. Locate the following points on the phase diagram requested in Exercise S11-15:

(a) triple point
(b) critical point
(c) normal melting point
(d) normal boiling point

S11-17. Sketch, again from memory, the following transitions on the phase diagram used in Exercises S11-15 and S11-16:

(a) freezing
(b) melting
(c) condensation
(d) vaporization
(e) deposition
(f) sublimation

Represent each transition as both an isothermal and an isobaric process.

S11-18. Sketch the phase diagram for a substance that is more dense in its solid form than in its liquid form.

S11-19. Sketch the phase diagram for a substance that undergoes an increase in molar volume upon freezing.

S11-20. The vapor pressure for substance 1 varies sharply with temperature, whereas for substance 2 the variation is far less pronounced. Sketch the phase diagram for each substance.

S11-21. Does the boiling temperature of water increase, decrease, or remain the same as the ambient pressure decreases? Show the effect on a phase diagram.

S11-22. Which is the ordered phase and which is the disordered phase in each pair?

(a) $CH_3CH_2OH(g)$, $CH_3CH_2OH(\ell)$

(b) $CH_3CH_2OH(s)$, $CH_3CH_2OH(\ell)$

(c) $NaCl(s)$, $NaCl(aq)$

(d) $H_2O(\ell)$ at 25°C, $H_2O(\ell)$ at 30°C

Chapter 12

Equilibrium—The Stable State

S12-1. Classify each of the following situations as either a dynamic equilibrium or a static equilibrium:

(a) A framed picture hangs motionless on a wall.
(b) A constant vapor pressure exists above an enclosed liquid.
(c) All portions of a drink are equally sweet.
(d) The value of $[H_3O^+]$ is constant in an acidic solution.

S12-2. Consider the process

$$A(aq) + B(aq) \rightleftarrows C(aq) + D(g) \qquad K = 0.001$$

Propose a way to achieve 100% conversion of reactants into products, despite the relative smallness of the equilibrium constant.

S12-3. Set up, but do not solve, the algebraic expression needed to establish the equilibrium pressure of each component in the reaction below:

$$2A(g) + 3B(g) \rightleftarrows C(g) + 4D(g) \qquad K = 15$$

Assume the following values for the initial pressures:

$$P_{A,0} = 10 \text{ atm} \qquad P_{B,0} = 20 \text{ atm} \qquad P_{C,0} = P_{D,0} = 0$$

S12-4. Consider the process below:

$$A(g) + B(g) \rightleftarrows 3C(s) + D(g) \qquad K = 0.0001$$

After equilibrium is established, the volume of the system is suddenly tripled. Does this change in volume affect the equilibrium? If so, then in which direction does the reaction shift—toward products or toward reactants?

S12-5. Assume that $K = 2$ for the following process:

$$A(g) \rightleftarrows 2B(g)$$

(a) If $P_A = 1$ atm at equilibrium, what is the corresponding value of P_B? **(b)** Once this first equilibrium is attained, suppose that the volume of the system is halved and component B is removed completely. Calculate P_A and P_B at the new equilibrium.

S12-6. Say that the reaction

$$A(g) + B(g) \rightleftarrows C(g)$$

comes to equilibrium with final pressures as given below:

$$P_A = 0.50 \text{ atm} \qquad P_B = 0.50 \text{ atm} \qquad P_C = 1.0 \text{ atm}$$

(a) Calculate the equilibrium constant. **(b)** Three hours after the attainment of this first equilibrium, the product C is completely removed from the system. Calculate the values of P_A, P_B, and P_C after equilibrium is reestablished.

S12-7. Repeat part (b) of the preceding exercise, but with this one modification: The product C is removed from the system only 1 μs after the first equilibrium is reached, not 3 hours as specified earlier. Are the results affected in any way?

S12-8. Assume that P_D remains constant at 1 atm after the following reaction comes to equilibrium at the stated temperature:

$$A(g) + B(g) \rightleftarrows C(g) + D(g) \qquad K(500°C) = 100$$

Do you have sufficient information to determine the equilibrium pressures of all the other species? If yes, do so. If no, explain why the problem is poorly posed.

S12-9. Reconsider the hypothetical process introduced in the preceding exercise:

$$A(g) + B(g) \rightleftarrows C(g) + D(g) \qquad K(500°C) = 100$$

Given the initial pressures stated below,

$$P_{A,0} = P_{B,0} = 10 \text{ atm} \qquad P_{C,0} = P_{D,0} = 0$$

do you have sufficient information to determine the equilibrium pressures of all components? If yes, do so. If no, explain why the problem is poorly posed.

S12-10. Suppose that [A] = [B] = 1 *M* when the following reaction comes to equilibrium:

$$A(aq) \rightleftarrows B(aq) \qquad K = 1$$

Propose two different—but equally possible—sets of initial concentrations that will produce exactly the same concentrations at equilibrium.

S12-11. The equilibrium constant for the reaction

$$HF(aq) + H_2O(\ell) \rightleftarrows H_3O^+(aq) + F^-(aq)$$

is given by the expression

$$K = \frac{[H_3O^+][F^-]}{[HF]}$$

Assume that 5.00 grams of hydrofluoric acid are initially dissolved in 1.00 L of water at 25°C. If $K(25°C) = 6.8 \times 10^{-4}$, what are the concentrations of hydronium and fluoride ions at equilibrium?

S12-12. Repeat the preceding exercise, but this time use an expression that explicitly includes the concentration of water, a pure liquid:

$$K' = \frac{[H_3O^+][F^-]}{[HF][H_2O]}$$

Are the results changed in any way?

S12-13. The equilibrium constant for the reaction

$$Ag_2CO_3(s) \rightleftarrows 2Ag^+(aq) + CO_3^{2-}(aq)$$

is $K(25°C) = 8.5 \times 10^{-12}$. Calculate the equilibrium concentration of each ion in an aqueous solution of silver carbonate at 25°C.

S12-14. The equilibrium constant for the reaction

$$AgI(s) \rightleftarrows Ag^+(aq) + I^-(aq)$$

is $K(25°C) = 8.3 \times 10^{-17}$. **(a)** How many grams of silver iodide can be dissolved in 100 mL H_2O at 25°C? **(b)** How many grams can be dissolved in 100 L?

S12-15. Suppose that AgI is added to a 1.00 M solution of AgNO$_3$. At equilibrium, do you expect [I$^-$] to be higher than, lower than, or the same as its value in a solution of AgI in pure water?

S12-16. Give the numerical value of K for the process

$$H_2O(g) \rightleftarrows H_2O(\ell)$$

at the following temperatures:

(a) 0°C   **(b)** 25°C   **(c)** 40°C   **(d)** 80°C

Relevant information may be found in Appendix C.

S12-17. Does the equilibrium constant for the process

$$C_6H_6(\ell) \rightleftarrows C_6H_6(g)$$

increase, decrease, or remain the same as the temperature is increased?

S12-18. Does the equilibrium constant for the process

$$H_2O(g) \rightleftarrows H_2O(s)$$

increase, decrease, or remain the same as the temperature is increased?

S12-19. Suppose that the process

$$CH_3CH_2OH(\ell) \rightleftarrows CH_3CH_2OH(g)$$

has attained equilibrium. **(a)** Describe what will happen if the volume of the enclosed system is suddenly increased. **(b)** Does the equilibrium constant increase, decrease, or remain the same as the volume is increased? **(c)** What will happen if the system is left open to the atmosphere?

S12-20. The conversion of graphite into diamond,

$$C(s, graphite) \rightleftarrows C(s, diamond)$$

proceeds on a geological time scale—a very long time. What can you say about the magnitude of K?

S12-21. The dissolution of sucrose in water,

$$C_{12}H_{22}O_{11}(s) \rightleftarrows C_{12}H_{22}O_{11}(aq)$$

proceeds in a matter of seconds. What can you say about the magnitude of K?

S12-22. A heap of NaCl remains undissolved in water until an experimenter vigorously stirs the solution. Does K increase, decrease, or remain the same as a result of the stirring?

S12-23. Our experimenter continues to dissolve NaCl in water until finally no more solid will dissolve, whereupon the additional sodium chloride falls to the bottom and resists even the most determined stirring. Does the position of equilibrium for the reaction

$$NaCl(s) \rightleftarrows Na^+(aq) + Cl^-(aq)$$

shift in any way as the excess solid accumulates? Is it legitimate to invoke Le Châteliers's principle?

Chapter 13

Energy, Heat, and Chemical Change

S13-1. Classify each property as extensive or intensive:

   (a) thermal conductivity   **(b)** melting point   **(c)** triple point   **(d)** color

Assume that the system is sufficiently large to behave in bulk.

S13-2. Which of the following quantities are functions of state? Which are functions of path?

$$\textbf{(a) } H - PV \qquad \textbf{(b) } E - TS \qquad \textbf{(c) } c_V \Delta T \qquad \textbf{(d) } \tfrac{3}{2} R \Delta T$$

S13-3. A fixed quantity of gaseous CO_2 (1 mol), initially at STP, is subjected to a rapid cycle of expansion and compression. When the process is over, the temperature is 273 K and the volume is 22.4 L. Assume that the gas behaves ideally. **(a)** Calculate ΔE. **(b)** Calculate ΔH. **(c)** Calculate ΔS.

S13-4. How much work is done when an ideal gas expands from 1.00 L to 3.00 L under a constant external pressure of 2.00 atm?

S13-5. Suppose that an ideal gas is confined to an insulating vessel, unable to exchange heat with its surroundings. Assume further that the internal pressure is equal to the external pressure. By itself, unassisted, is the gas able to expand?

S13-6. A sample of helium gas at STP is confined by a piston to a volume of 10.0 L. System and surroundings are both at the same temperature, able to exchange heat freely. **(a)** When the pressure is doubled, the volume of gas falls to 5.0 L. Calculate ΔE. **(b)** Is there a flow of heat? If so, in which direction does it proceed—from system to surroundings, or from surroundings to system?

S13-7. Consider the following two states of an ideal gas, both at STP:

State 1: $n = 1.00$ mol   State 2: $n = 2.00$ mol

(a) Calculate $\Delta E = E_2 - E_1$. **(b)** Calculate $\Delta H = H_2 - H_1$.

S13-8. The temperature of argon gas in a closed system (1.00 mol) falls from 300 K to 200 K at constant volume. **(a)** In which direction does the heat flow? **(b)** Do you have enough information to compute w? If yes, do so. If no, explain why. **(c)** Do you have enough information to compute q, ΔE, and ΔH? If yes, do so. If no, explain why.

S13-9. A system of neon gas at STP is confined to a rigid volume of 224.1 L. Calculate ΔE and ΔT after the gas absorbs a heat flow of 10.0 kJ.

S13-10. The temperature of helium gas in a rigid vessel rises by 10.0 K after the system absorbs 5.00 kJ of heat. Calculate the mass of helium in the vessel.

S13-11. A beaker of water containing 1.00 mol, open to the atmosphere at 20°C, is brought into contact with an object at a lower temperature. If the temperature of the water falls by 10.15°C, what is the corresponding change in enthalpy? Relevant data may be found in Appendix C.

S13-12. Consult Appendix C for any data required to answer the following questions. **(a)** How many joules are needed to raise the temperature of 3.17 grams of potassium metal by 15.0°C at constant pressure? **(b)** How many joules are needed to bring about the same change in temperature for 3.17 grams of lithium?

S13-13. What additional information about the system do you need in order to compute $\Delta H°$ for the indicated reaction?

$$2H_2 + O_2 \rightarrow 2H_2O$$

S13-14. Use the data in Appendix C to calculate $\Delta H°$ for each of the reactions below:

(a) $4NH_3(g) + 5O_2(g) \rightarrow 4NO(g) + 6H_2O(g)$
(b) $4NH_3(g) + 3O_2(g) \rightarrow 2N_2(g) + 6H_2O(g)$
(c) $2NH_3(g) + 2O_2(g) \rightarrow N_2O(g) + 3H_2O(g)$

S13-15. Use the data in Appendix C to calculate the heat of combustion per mole of liquid octane at 1 atm and 25°C.

S13-16. How much heat must be supplied to vaporize 100.0 grams of metallic silver at 1 atm and 25°C? Consult Appendix C as needed.

S13-17. Again, use Appendix C where needed. **(a)** Calculate the change in enthalpy when a 456-g sample of benzene condenses from gas to liquid at 1 atm and 25°C. **(b)** Assume that the heat evolved is captured entirely by 5.00 L of water at 25.00°C. What is the final temperature of the water?

S13-18. Similar. **(a)** Calculate the change in enthalpy when a 755-mL sample of ethanol vaporizes at 1 atm and 25°C. The density of ethanol is 0.7873 g mL^{-1}. **(b)** Assume that the heat required for this vaporization is withdrawn from 1000 moles of helium at 25.0°C, the transfer occurring at constant pressure. What is the final temperature of the helium?

S13-19. Suppose that a typographical error has caused all the enthalpies of formation in Table C-16 to be exactly 1.0 kJ mol^{-1} higher than intended. Would our calculations be affected in any way? If so, how?

S13-20. How can we say that heat (q) is not a function of state, but that enthalpy ($H \equiv q_P$) *is* a function of state? After all, isn't enthalpy equivalent to the heat exchanged under constant pressure?

S13-21. The dissolution of potassium chloride in water is an endothermic reaction. Would you expect a beaker of water to grow warm to the touch or cold to the touch as KCl is stirred into solution?

S13-22. Say that the dissolution of a certain salt in water is an exothermic reaction. Would you expect more material or less material to dissolve as the temperature is increased?

S13-23. Show, algebraically, that ΔH for the reaction

$$a\text{A} + b\text{B} \rightarrow c\text{C} + d\text{D}$$

is equal in magnitude and opposite in sign to ΔH for the reaction

$$c\text{C} + d\text{D} \rightarrow a\text{A} + b\text{B}$$

S13-24. Calculate the enthalpy of sublimation for iodine at 1 atm and 25°C.

S13-25. Consider the following process:

$$2\text{H(g)} \rightarrow \text{H}_2\text{(g)} \qquad \Delta H° = -436 \text{ kJ}$$

What is the standard heat of formation for H_2(g)? Does the reaction enthalpy specified above imply somehow that $\Delta H_f° = -436$ kJ mol^{-1} for molecular hydrogen—in contradiction to our convention concerning elements in their standard states?

Chapter 14

Free Energy and the Direction of Change

S14-1. Suppose that a coin has three sides: heads (H), tails (T), and feet (F). **(a)** How many outcomes are possible when the coin is tossed twice? List them. **(b)** Which outcome or outcomes are more probable than the rest?

S14-2. System 1 absorbs 100 joules of heat reversibly at a temperature of 5 K. System 2 also absorbs 100 joules reversibly, but at a temperature of 500 K. **(a)** For each system, state whether the change in entropy is positive, negative, or zero. **(b)** Which system realizes the greater change in entropy?

S14-3. Say that a system of gas exchanges thermal energy reversibly with a heat reservoir at 100 K. **(a)** If the entropy of the system decreases by 10 J K^{-1}, what quantity of heat is exchanged? **(b)** In which direction does the heat flow—from the reservoir to the gas, or from the gas to the reservoir?

S14-4. One mole of gas absorbs 100 joules of heat reversibly from a reservoir at 25°C. Calculate the change in entropy for the gas.

S14-5. A frictionless system, thermally isolated from its surroundings, is subjected to a reversible mechanical process. The work done is equal to 1 kJ. **(a)** Calculate ΔE. **(b)** Calculate ΔS.

S14-6. A system undergoes a series of reversible changes in which heat flows neither in nor out. **(a)** Does the entropy of the system increase, decrease, or remain the same? **(b)** Does the entropy of the surroundings increase, decrease, or remain the same? **(c)** Does the entropy of the universe increase, decrease, or remain the same?

S14-7. A system spontaneously undergoes an irreversible endothermic reaction. **(a)** Are you able to determine whether the change in entropy for the system is positive, negative or zero? **(b)** If yes, write a symbolic expression for ΔS in terms of other thermodynamic variables. If no, explain why not.

S14-8. As above, a system spontaneously undergoes an irreversible endothermic reaction. **(a)** Does the entropy of the surroundings increase, decrease, or remain the same? **(b)** Does the entropy of the universe increase, decrease, or remain the same?

S14-9. A system spontaneously undergoes an irreversible exothermic reaction. **(a)** Does the entropy of the surroundings increase, decrease, or remain the same? **(b)** Does the entropy of the universe increase, decrease, or remain the same?

S14-10. A confined gas undergoes a reversible change in internal energy, ΔE_{rev}, at constant volume. What is the corresponding change in entropy for the surroundings? Assume that system and surroundings are in thermal equilibrium.

S14-11. If pressure and temperature are constant, then the quantity $-\Delta G/T$ registers the change in entropy for system and surroundings combined—the universe. Recall, in particular, that a change in the Gibbs free energy under these conditions is defined as

$$\Delta G = \Delta H - T\,\Delta S$$

Now consider a change in the *Helmholtz* free energy, stated provisionally as

$$\Delta A = \Delta E - T\,\Delta S$$

Show that the quantity $-\Delta A/T$ registers the global change in entropy for a transformation undergone by a system at constant volume and temperature. Assume that system and surroundings are in thermal equilibrium.

S14-12. Assume that $\Delta H^{\circ} > 0$ and $K > 1$ for the hypothetical process

$$A + B \rightleftarrows C + D$$

at a certain temperature T. What is the algebraic sign of ΔS°?

S14-13. Assume that $\Delta H^{\circ} < 0$ and $K < 1$ for the hypothetical process

$$A \rightleftarrows B$$

at a certain temperature T. Which species has the higher standard entropy, A or B?

S14-14. Assume that $\Delta H^{\circ} = 100$ kJ and $\Delta S^{\circ} = 100$ J K^{-1} for the hypothetical process

$$A + 2B + 5C \rightleftarrows 3D + 4E$$

(a) Calculate the equilibrium constant at 1000 K. **(b)** Over what range of temperature, if any, is the reaction spontaneous?

S14-15. Which species in each pair has the higher value of ΔG_f° at 298 K?

        (a) $H_2(g)$, $H_2(\ell)$ **(b)** $H_2O(g)$, $H_2O(\ell)$ **(c)** $Na(\ell)$, $Na(s)$

Try to answer without consulting Appendix C.

S14-16. Again, but consider a higher temperature: Which species has the greater value of ΔG_f° at 450 K?

        (a) $H_2(g)$, $H_2(\ell)$ **(b)** $H_2O(g)$, $H_2O(\ell)$ **(c)** $Na(\ell)$, $Na(s)$

Try to answer without consulting Appendix C. Note from Table C-9, however, that the melting point of sodium is 97.7°C.

S14-17. (a) Calculate ΔH°, ΔS°, and ΔG° for the reaction below:

$$C_6H_{12}O_6(s) + 6O_2(g) \rightleftarrows 6CO_2(g) + 6H_2O(\ell)$$

(b) Over what range of temperature is the process spontaneous?

S14-18. Calculate ΔG_f° for $N_2O_5(g)$ given the information below:

$$N_2O_5(g) + H_2O(\ell) \rightleftarrows 2HNO_3(\ell) \qquad \Delta G^\circ = -39.3 \text{ kJ}$$

Other relevant data may be found in Appendix C.

S14-19. Calculate ΔH°, ΔS°, ΔG°, and K for the reaction below:

$$N_2(g) + 3O_2(g) + H_2(g) \rightleftarrows 2HNO_3(\ell)$$

Assume a temperature of 298 K.

S14-20. Use the data in Appendix C to estimate the boiling temperature of carbon tetrachloride, CCl_4. (The published boiling point of CCl_4 is 76.8°C, or 350 K.)

S14-21. The normal boiling point of methanol (CH_3OH) is 64.6°C, and the standard entropy of vaporization is 112.9 J mol^{-1} K^{-1}. **(a)** Without consulting Appendix C, estimate ΔH° for the reaction below:

$$CH_3OH(g) \rightleftarrows CH_3OH(\ell)$$

(b) Calculate $\Delta H°$ directly from the appropriate data in Appendix C. Compare the two results.

S14-22. Thermodynamic data for substances A, B, C, and D are tabulated below:

$$\Delta G_f° \text{ (kJ mol}^{-1}\text{)}$$

A	50
B	100
C	200
D	150

(a) Calculate the value of ΔG (not $\Delta G°$) when the reaction

$$A + 3B \rightleftarrows 2C + D$$

comes to equilibrium at 25°C. **(b)** Is the equilibrium constant greater than 1, less than 1, or equal to 1 at 25°C?

S14-23. Use the hypothetical formation data from the preceding exercise to solve the following problem: **(a)** Calculate the value of ΔG when the reaction

$$4A + 6B \rightleftarrows C + D$$

comes to equilibrium at 25°C. **(b)** Is the equilibrium constant greater than 1, less than 1, or equal to 1 at 25°C? Compare the results with those of the preceding exercise.

Chapter 15

Making Accommodations—Solubility and Molecular Recognition

S15-1. Explain the difference between the *solubility* and the *solubility-product constant* of a substance.

S15-2. One mole of X requires 30 seconds to dissolve in 1 liter of water at 10°C. One mole of Y requires 300 seconds to dissolve in the same amount of water at the same temperature. At 90°C, however, the times are equal: one mole of X dissolves in 3 seconds, and 1 mole of Y dissolves in 3 seconds as well. Note further that no undissolved solute is evident in any of the solutions considered. **(a)** Can you say whether the reactions

$$X(s) \rightleftarrows X(aq) \quad \text{and} \quad Y(s) \rightleftarrows Y(aq)$$

are exothermic or endothermic? If yes, do so. If no, explain why not. **(b)** Can you determine whether the solubility of X is greater than, less than, or equal to the solubility of Y at each temperature? If yes, do so. If no, explain why not. **(c)** Can you calculate K_{sp} for X and Y at each temperature? If yes, do so. If no, explain why not.

S15-3. Suppose that a saturated solution of X (volume = 1 L) is produced after 10 s of rapid stirring at 25°C. Thermodynamic data are given below:

	ΔH_f° (kJ mol^{-1})	ΔG_f° (kJ mol^{-1})
X(s)	120	100
X(aq)	140	80

(a) Can you determine whether the solubility of X at 50°C will be greater than, less than, or equal to the solubility at 25°C? If yes, do so. If no, explain why not. **(b)** Can you determine whether the saturation point at 50°C will be reached in 10 s, more than 10 s, or less than 10 s? If yes, do so. If no, explain why not.

S15-4. Consider a generic ionic compound A_jB_k, where A denotes a cation with charge p and B denotes an anion with charge $-q$. **(a)** Determine the algebraic relationship between p and q. **(b)** Write a symbolic algebraic expression for the solubility-product constant of A_jB_k. **(c)** Write an equation that gives the molar solubility of A_jB_k in terms of only j, k, and the numerical value of K_{sp}.

S15-5. Assume that the numerical value of K_{sp} at some particular temperature is the same for compounds AB_2 and AB_3. **(a)** Calculate the ratio of solubilities. **(b)** Is the concentration of cations in a saturated solution of AB_2 necessarily equal to the concentration of cations in a saturated solution of AB_3? **(c)** For saturated solutions of both AB_2 and AB_3, express the concentration of anions in terms of K_{sp}.

S15-6. The solubility of $Mg(OH)_2$ is approximately 0.0065 g L^{-1} at 25°C. Calculate K_{sp}.

S15-7. The solubility-product constant for SrF_2 is $K_{sp}(25°C) = 4.3 \times 10^{-9}$. Calculate the solubility of SrF_2 in g L^{-1} at 25°C.

S15-8. Rank the following ionic solids in order of increasing molar solubility in water:

$$Ca(NO_3)_2, \quad Hg_2I_2, \quad MgF_2, \quad ZnCO_3$$

Assume a temperature of 25°C. Relevant information may be found in Appendix C.

S15-9. The solubility-product constant for $AgCl$ is $K_{sp}(25°C) = 1.8 \times 10^{-10}$. **(a)** What minimum volume of solution is required to dissolve 1.0 g $AgCl$ in water at 25°C? **(b)** What minimum volume of solution is required to dissolve 1.0×10^2 g?

S15-10. The solubility-product constant for $PbCl_2$ is $K_{sp}(25°C) = 1.6 \times 10^{-5}$. **(a)** What minimum volume of solution is required to dissolve 1.0 g $PbCl_2$ in water at 25°C? **(b)** What minimum volume of solution is required to dissolve 1.0×10^2 g?

S15-11. Assume that x grams of $MgCO_3$ can be dissolved in 100 milliliters of pure water. Will the mass of $MgCO_3$ that can be dissolved in 100 milliliters of 1.00 M Na_2SO_4 be greater than, less than, or the same as x?

S15-12. Assume, again, that x grams of $MgCO_3$ can be dissolved in 100 milliliters of pure water. Will the mass of $MgCO_3$ that can be dissolved in 100 milliliters of 1.00 M $MgSO_4$ be greater than, less than, or the same as x?

S15-13. Imagine that a saturated solution of $Ag^+(aq)$ and $Cl^-(aq)$ coexists in equilibrium with an infinitely large pile of $AgCl(s)$. The temperature is 25°C. **(a)** What happens as additional water drips into the solution? **(b)** Does the concentration of $AgCl(aq)$ increase, decrease, or remain the same as water is added?

S15-14. Consider a variation of the preceding exercise: A saturated solution of AgCl(aq) comes to equilibrium not with an infinitely large pile of AgCl(s), but rather with just a single grain. Again, the temperature is 25°C. **(a)** Is the equilibrium concentration of AgCl(aq) in this new situation greater than, less than, or equal to its previous value? **(b)** What happens as additional water drips into the solution? **(c)** Does the concentration of AgCl(aq) increase, decrease, or remain the same as more water is added?

S15-15. Suppose that a saturated solution abruptly loses 25% of its volume owing to evaporation of the solvent. **(a)** If a precipitate does not form immediately, does a heterogeneous equilibrium exist in the solution? **(b)** In the absence of precipitation, is the numerical value of Q (the reaction quotient) greater than, less than, or equal to K_{sp}? **(c)** Does the numerical value of K_{sp} change as the volume changes? **(d)** Is the *equilibrium* concentration of solute in the smaller volume greater than, less than, or equal to its value in the larger volume?

S15-16. Consider a specific numerical example to illustrate the idea developed in the previous exercise: A saturated solution of AgOH ($K_{sp} = 1.8 \times 10^{-8}$ at 25°C) loses 25% of its volume owing to evaporation. Calculate the equilibrium concentrations of Ag^+ and OH^- before and after evaporation.

S15-17. Study the thermodynamic data given below for two hypothetical equilibria:

REACTION	$\Delta H°$ (kJ mol^{-1})	$\Delta S°$ (J mol^{-1} K^{-1})
AB(s) $\rightleftarrows$ A$^+$(aq) + B$^-$(aq)	50	100
CD(s) $\rightleftarrows$ C$^+$(aq) + D$^-$(aq)	−50	100

(a) Which of the two ionic solutes—AB or CD—is more soluble at 25°C? **(b)** Does the solubility of AB increase, decrease, or remain the same at temperatures greater than 25°C? **(c)** Similarly, for CD: Does the solubility increase, decrease, or remain the same at higher temperatures? Account for the direction of change.

S15-18. Another pair of hypothetical equilibria:

REACTION	$\Delta H°$ (kJ mol^{-1})	$\Delta S°$ (J mol^{-1} K^{-1})
AB(s) $\rightleftarrows$ A$^+$(aq) + B$^-$(aq)	50	100
CD(s) $\rightleftarrows$ C$^+$(aq) + D$^-$(aq)	−50	−300

(a) Which of the two ionic solutes—AB or CD—is more soluble at 25°C? **(b)** Does the solubility of AB increase, decrease, or remain the same at temperatures greater than 25°C? **(c)** Does the solubility of CD increase, decrease, or remain the same at higher temperatures? Account for the direction of change.

S15-19. Study the thermodynamic data given below for two hypothetical equilibria:

REACTION	$\Delta H°$ (kJ mol^{-1})	$\Delta S°$ (J mol^{-1} K^{-1})
AB(s) $\rightleftarrows$ A$^+$(aq) + B$^-$(aq)	50	100
CD(s) $\rightleftarrows$ C$^+$(aq) + D$^-$(aq)	5	-100

(a) Calculate the ratio K_{sp}(AB)/K_{sp}(CD) at 25°C. **(b)** Calculate the ratio K_{sp}(AB)/K_{sp}(CD) at 4°C. **(c)** Account for the different ratios at the different temperatures.

S15-20. Use the formation data in Appendix C to calculate K_{sp} for Li_2CO_3 at 25°C.

S15-21. Would you expect K_{sp} for $Ca(NO_3)_2$ to be greater than 1, less than 1, or equal to 1? Relevant data may be found in Appendix C.

S15-22. Two halves of a glass vessel are separated by a semipermeable membrane open only to the passage of water molecules, not sodium ions. On the left-hand side, the concentration of Na^+ has the value x. On the right-hand side, an experimenter adjusts the initial concentration of Na^+ to the value $x/2$. The volume of liquid in each half is 1 L. **(a)** Are the two solutions in osmotic equilibrium? **(b)** If not, in which direction will the water flow—from left to right or from right to left? **(c)** What will be the final concentration of Na^+ in each half? **(d)** Is the flow spontaneous or nonspontaneous?

S15-23. Now, what's the difference between a glass vessel and a living cell? Pursue the following analogy: Inside a cell, the concentration of Na^+ has a certain value x. In the water outside the cell, the concentration of Na^+ has the value $x/2$. The intracellular and extracellular regions are separated by a semipermeable membrane through which H_2O molecules (although not sodium ions) can pass. Nevertheless, the concentration of Na^+ inside the cell remains x—unchanged—and the concentration of Na^+ outside the cell remains $x/2$. **(a)** Is the cell in osmotic equilibrium with the external world? If not, explain how the living system can resist nature's powerful tendency to "smooth away the differences." **(b)** Does a living cell violate, in any way, the second law of thermodynamics? If it does, what shall we make of our so-called *law*? If the cell does not violate any law of nature, then how does it manage to conduct its chemical activities legally?

S15-24. What information do you need in order to determine the molar mass of a solute by measurement of osmotic pressure?

S15-25. **(a)** Does the osmotic pressure of a solution increase, decrease, or remain the same as the temperature is increased? **(b)** If solution 1 is isotonic with solution 2 at a certain temperature T, do the two solutions stay isotonic as the temperature is increased? **(c)** What happens if solution 1 (originally isotonic with solution 2) loses half its volume owing to evaporation? Do the two solutions remain isotonic? If not, does solution 1 become hypotonic or hypertonic relative to solution 2?

S15-26. An aqueous solution of sodium chloride (call it solution 1) contains 2.000 g NaCl in a total volume of 100.0 mL. **(a)** How many grams of $Ce_2(SO_4)_3$ must be dissolved in a total volume of 200.0 mL to produce a second solution (call it solution 2) isotonic with solution 1 at 5°C? **(b)** Which piece of information given in part (a) is superfluous to this exercise?

S15-27. A mixture of nitrogen and oxygen gases is in contact with liquid water at 20°C. Henry's law constants are given below:

$$N_2(aq) \rightleftarrows N_2(g) \qquad K(20°C) = 1.4 \times 10^3 \text{ atm L mol}^{-1}$$
$$O_2(aq) \rightleftarrows O_2(g) \qquad K(20°C) = 7.2 \times 10^2 \text{ atm L mol}^{-1}$$

The partial pressure of N_2 is 50 atm, and the partial pressure of O_2 is 10 atm. **(a)** Calculate the solubility of N_2 under the stated conditions. **(b)** Calculate the solubility of N_2 after the partial pressure of O_2 is increased to 100 atm.

S15-28. $AgNO_3$ is strongly soluble in water. AgI is nearly insoluble, with a K_{sp} less than 10^{-16}. At equilibrium, state whether the residual free energy (ΔG) for the dissolution of silver nitrate,

$$AgNO_3(s) \rightleftarrows Ag^+(aq) + NO_3^-(aq)$$

is greater than, less than, or equal to the residual free energy for the dissolution of silver iodide:

$$AgI(s) \rightleftarrows Ag^+(aq) + I^-(aq)$$

Be sure to distinguish ΔG from $\Delta G°$.

Chapter 16

Acids and Bases

S16-1. Listed below are four statements concerning an aqueous solution of HCl at 25°C. Which one is true?

(a) The pH is always less than 7.
(b) The pH is always greater than 7.
(c) The pH is always greater than 0.
(d) The pH is usually 1.

S16-2. An aqueous solution contains 0.000001 mol HCl. Which of the following statements is true?

(a) The pH of the solution is 6.
(b) The pH of the solution is −6.
(c) The pH of the solution lies between 6 and 7.
(d) There is insufficient information to determine the pH.

S16-3. Beaker 1 contains a solution of hydrochloric acid in water. Beaker 2 contains a solution of acetic acid in water, the same volume as in beaker 1. Which of the following statements is true?

(a) The pH of solution 1 is always higher than the pH of solution 2.
(b) The pH of solution 1 is always lower than the pH of solution 2.
(c) The pH of solution 1 is always equal to the pH of solution 2.
(d) The pH of solution 1 may be higher than, lower than, or equal to the pH of solution 2.

S16-4. The initial concentration of HCl in a certain solution is 10^{-8} M. Which of the following statements is true?

(a) The pH at 25°C is exactly 6.
(b) The pH at 25°C is slightly less than 7.
(c) The pH at 25°C is exactly 7.
(d) The pH at 25°C is slightly greater than 7.
(e) The pH at 25°C is exactly 8.

S16-5. The equilibrium constant for the autoionization of water is 1.0×10^{-14} at 25°C:

$$H_2O(\ell) + H_2O(\ell) \rightleftarrows H_3O^+(aq) + OH^-(aq) \qquad K_w(25°C) = 1.0 \times 10^{-14}$$

Which of the following statements is true for a strongly *acidic* solution at 25°C?

(a) The numerical value of K_w is greater than 10^{-14}.
(b) The numerical value of K_w is less than 10^{-14}.
(c) The numerical value of K_w is equal to 10^{-14}.
(d) There is insufficient information to determine the value of K_w in the presence of acid.

S16-6. Consider again the autoionization of water:

$$H_2O(\ell) + H_2O(\ell) \rightleftarrows H_3O^+(aq) + OH^-(aq)$$

Which of the following statements is true?

(a) The equilibrium shifts to the left in the presence of strong acid.
(b) The equilibrium shifts to the right in the presence of strong acid.
(c) The equilibrium is unaffected by the presence of strong acid.
(d) There is insufficient information to determine the effect of strong acid on the equilibrium.

S16-7. The pH of pure water at 10°C is 7.26. **(a)** Determine the value of K_w at this temperature. **(b)** Is water acidic, basic, or neutral at 10°C?

S16-8. (a) Use the appropriate data in Appendix C to compute the pH of pure water at 50°C. **(b)** Compute the corresponding pOH. **(c)** Is water acidic, basic, or neutral at 50°C?

S16-9. Is it possible for an aqueous solution of NaOH at 25°C to have a pH less than 7? If yes, give an example of such a solution. If no, explain why not.

S16-10. An aqueous solution contains 45.0 g NaCl dissolved in a total volume of 450.0 mL. **(a)** Calculate the pH at 25°C. **(b)** What happens to the pH when the solution is diluted by a factor of 10?

S16-11. For each of the following salts, write a neutralization reaction able to produce the given compound:

      (a) KIO_4 **(b)** $Ca(HCOO)_2$ **(c)** NH_4CH_3COO **(d)** BaI_2

S16-12. Rank the following acids in order of increasing strength:

$$H_3AsO_4, \ H_2CO_3, \ HClO_3, \ CF_3COOH$$

Consult Appendix C as needed.

S16-13. Rank the following conjugate bases in order of increasing strength:

$$HCOO^-, \ F^-, \ BrO^-, \ SCN^-$$

Consult Appendix C as needed.

S16-14. For each of the following salts, state whether the pH of an aqueous solution at 25°C is greater than, less than, or equal to 7—or, perhaps, impossible to establish without further information:

      (a) $NaCH_3CH_2COO$ **(b)** KBr **(c)** NH_4Cl **(d)** $NaNO_3$

Consult Appendix C as needed.

S16-15. Bromoacetic acid ($CH_2BrCOOH$) has a pK_a value of 2.70. Given an initial concentration $[CH_2BrCOOH]_0$, calculate the pH and the ratio $[CH_2BrCOO^-]/[CH_2BrCOOH]$ at equilibrium:

(a) $[CH_2BrCOOH]_0 = 1.000 \ M$
(b) $[CH_2BrCOOH]_0 = 0.500 \ M$
(c) $[CH_2BrCOOH]_0 = 0.100 \ M$
(d) $[CH_2BrCOOH]_0 = 0.010 \ M$

S16-16. A buffer solution is prepared as a mixture of bromoacetic acid ($pK_a = 2.70$) and its conjugate base. The pH is 3.00. **(a)** In what proportions must the two components be mixed in order to realize the stated pH? **(b)** Suppose that a certain amount of NaOH is added to the buffer solution. Do you have sufficient information to determine the change in pH? If yes, do so. If no, what further information do you need?

S16-17. A mixture contains 0.00100 M $CH_2BrCOOH$ and 0.00100 M CH_2BrCOO^-. The ionization constant of bromoacetic acid is $K_a(25°C) = 0.0020$. **(a)** Do you expect this system to act as an effective buffer? **(b)** Calculate the pH. Is it appropriate to apply the Henderson-Hasselbalch equation here?

S16-18. Given an initial concentration of formic acid, $[HCOOH]_0$, calculate the pH and the ratio $[HCOO^-]/[HCOOH]$ at equilibrium:

(a) $[HCOOH]_0 = 1.000\ M$

(b) $[HCOOH]_0 = 0.500\ M$

(c) $[HCOOH]_0 = 0.100\ M$

(d) $[HCOOH]_0 = 0.010\ M$

The ionization constant of HCOOH is $K_a(25°C) = 1.8 \times 10^{-4}$.

S16-19. A buffer solution is prepared from formic acid (HCOOH) and sodium formate (NaHCOO). The concentration of acid is $1.000\ M$ in a volume of $1.000\ L$, and the pH of the solution is 3.80. **(a)** Calculate the formate concentration. **(b)** Calculate the pH of the buffer solution after addition of 0.100 mol NaOH. **(c)** Recalculate the pH of the original buffer solution after addition of 0.100 mol HCl.

S16-20. A buffer solution contains 5.75 g HCOOH and 6.80 g NaHCOO dissolved in a total volume of 250. mL. **(a)** Calculate the pH. **(b)** What concentration of HCOOH will adjust the pH to 3.50? **(c)** What concentration of HCl, if added to the original buffer, will likewise adjust the pH to 3.50? Assume that the volume does not change.

S16-21. A buffer solution contains 1.000 mol HA and 1.000 mol A^- in a total volume of $1.000\ L$. Assume that the ionization constant for the reaction

$$HA(aq) + H_2O(\ell) \rightleftarrows A^-(aq) + H_3O^+(aq)$$

has the value $K_a = 1.00 \times 10^{-4}$. **(a)** Calculate the pH. **(b)** Recalculate the pH after addition of 0.001 mol HCl. **(c)** Increase the volume of the original buffer solution by a factor of 100, from $1.000\ L$ to $100.000\ L$. Recalculate the pH. **(d)** Add 0.1 mol HCl to the diluted buffer solution described in part (c). What is the pH now?

S16-22. Suppose that the ionization reaction of HA is exothermic. Will the pH of an aqueous solution of HA increase, decrease, or remain the same as the temperature is increased?

S16-23. Will the pH of an aqueous solution of HA increase, decrease, or remain the same as the pressure is increased?

S16-24. Will the pH of an aqueous solution of HA increase, decrease, or remain the same when a given amount of HA is dissolved in a larger volume of solvent?

S16-25. HNO_3 is a strong acid. HCN is a weak acid. At equilibrium, state whether the residual free energy (ΔG) for the ionization of nitric acid,

$$HNO_3(aq) + H_2O(\ell) \rightleftarrows NO_3^-(aq) + H_3O^+(aq)$$

is greater than, less than, or equal to the residual free energy for the ionization of hydrocyanic acid:

$$HCN(aq) + H_2O(\ell) \rightleftarrows CN^-(aq) + H_3O^+(aq)$$

Be sure to distinguish ΔG from $\Delta G°$.

Chapter 17

Chemistry and Electricity

S17-1. Which of the following half-reactions will probably have the higher reduction potential?

$$XeO_3(aq) + 6H^+(aq) + 6e^- \rightarrow Xe(g) + 3H_2O(\ell)$$

$$Cu^{2+}(aq) + 2e^- \rightarrow Cu(s)$$

Use general chemical arguments—not a table of electrode potentials—to make your assessment.

S17-2. Which of the following half-reactions will probably have the higher reduction potential?

$$O_3(g) + 2H^+(aq) + 2e^- \rightarrow O_2(g) + H_2O(\ell)$$

$$K^+(aq) + e^- \rightarrow K(s)$$

Use general chemical arguments to make your assessment.

S17-3. Which of the following metals is oxidized more readily than zinc?

(a) Ag **(b)** Na **(c)** Al **(d)** Ni

S17-4. By custom, the reduction of H^+ to H_2 is assigned a standard potential of zero volts:

$$2H^+(aq, 1\ M) + 2e^- \rightarrow H_2(g, 1\ atm) \qquad \mathcal{E}^{\circ}_{red} = 0\ V \ \text{(conventional)}$$

Suppose, instead, that we use the reduction of Cu^{2+} to establish a new zero point:

$$Cu^{2+}(aq, 1\ M) + 2e^- \rightarrow Cu(s) \qquad\qquad \mathcal{E}^\circ_{red} = 0\ V\ \text{(rescaled)}$$

(a) What voltage would be obtained for the reduction of Zn^{2+} to Zn in the rescaled system? If necessary, consult the relevant table in Appendix C. **(b)** What voltage would be obtained for the zinc–copper reaction written below?

$$Zn(s) + Cu^{2+}(aq) \rightarrow Zn^{2+}(aq) + Cu(s)$$

(c) Is the rescaled voltage different from the conventional voltage?

S17-5. Species A, which exists in the forms A(s) and $A^{p+}(aq)$, has a standard reduction potential of 1.00 V. Species B, which exists as B(s) and $B^{q+}(aq)$, has a standard reduction potential of 2.00 V. **(a)** Write a balanced equation that depicts a spontaneous redox reaction between A and B. **(b)** Which species is the oxidizing agent? Which is the reducing agent? **(c)** Calculate the standard cell potential.

S17-6. Consider the two half-reactions below:

$$A(s) \rightarrow A^+(aq) + e^- \qquad\qquad \mathcal{E}^\circ_{ox} = 1.00\ V$$

$$3e^- + B^{3+}(aq) \rightarrow B(s) \qquad\qquad \mathcal{E}^\circ_{red} = -1.00\ V$$

(a) Calculate the standard cell potential for the process

$$3A(s) + B^{3+}(aq) \rightarrow 3A^+(aq) + B(s)$$

(b) Calculate ΔG° and K (the equilibrium constant). Is the reaction spontaneous under standard conditions?

S17-7. Consider the following half-reactions:

$$A(s) \rightarrow A^+(aq) + e^- \qquad\qquad \mathcal{E}^\circ_{ox} = 2.00\ V$$

$$4e^- + B^{4+}(aq) \rightarrow B(s) \qquad\qquad \mathcal{E}^\circ_{red} = 1.00\ V$$

(a) For which *separate* process, reduction or oxidation, is ΔG° more negative? Treat each half-reaction in isolation, as written. **(b)** Calculate $\mathcal{E}^\circ$ for the coupled reaction below:

$$4A(s) + B^{4+}(aq) \rightarrow 4A^+(aq) + B(s)$$

(c) Suppose that $[A^+] = 1.1\ M$ and $[B^{4+}] = 2.0\ M$. Is the resulting cell potential greater than, less than, or equal to $\mathcal{E}^\circ$?

S17-8. Consider an arbitrary reaction for which the standard cell potential is 2.00 V:

$$A^+(aq) + B(g) \rightarrow A(g) + B^+(aq) \qquad \mathcal{E}^\circ = 2.00 \text{ V}$$

The voltage is subsequently measured at 25°C over a range of partial pressures, with concentrations fixed at 1 M. Which of the following statements is always true?

(a) The cell potential is 2.00 V when $P_A = 1$ atm.
(b) The cell potential is 2.00 V when $P_B = 1$ atm.
(c) The cell potential is 2.00 V when $P_A = P_B$.
(d) The cell potential is 2.00 V only when $P_A = P_B = 1$ atm.

S17-9. Again, consider the reaction specified below:

$$A^+(aq) + B(g) \rightarrow A(g) + B^+(aq) \qquad \mathcal{E}^\circ = 2.00 \text{ V}$$

The voltage is measured at 25°C over a range of concentrations, with all partial pressures fixed at 1 atm. Which of the following statements is always true?

(a) The cell potential is 2.00 V when $[A^+] = 1$ M.
(b) The cell potential is 2.00 V when $[B^+] = 1$ M.
(c) The cell potential is 2.00 V when $[A^+] = [B^+]$.
(d) The cell potential is 2.00 V only when $[A^+] = [B^+] = 1$ M.

S17-10. For the last time:

$$A^+(aq) + B(g) \rightarrow A(g) + B^+(aq) \qquad \mathcal{E}^\circ = 2.00 \text{ V}$$

An experimenter measures a cell potential of 1.99 V under the conditions stated below:

$$P_A = P_B = 1 \text{ atm} \qquad [A^+] = 1 \text{ } M \qquad T = 298.15 \text{ K}$$

(a) Is the value of $[B^+]$ greater than, less than, or equal to 1 M? **(b)** Is the reaction spontaneous in the forward direction?

S17-11. Use the formation data in Table C-16 to calculate the standard oxidation potential for the following reaction:

$$Ag(s) + Cl^-(aq) \rightarrow AgCl(s) + e^-$$

S17-12. (a) Without consulting Table C-21, use the formation data in Table C-16 to calculate the standard cell potential for the reaction below:

$$Cu(s) + Cl_2(g) \rightarrow Cu^{2+}(aq) + 2Cl^-(aq)$$

(b) Recalculate the standard voltage, this time using the reduction potentials in Table C-21.

S17-13. The reduction potentials tabulated in Appendix C were measured under standard conditions: pressures of 1 atm, concentrations of 1 M, temperatures of 25°C. Recalculate the reduction potential for the half-reaction

$$Zn^{2+}(aq) + 2e^- \rightarrow Zn(s)$$

under the conditions stated below:

	$[Zn^{2+}]$ (M)	P (atm)	T (K)
(a)	0.100	1.00	298.15
(b)	10.000	1.00	298.15
(c)	1.000	0.10	298.15
(d)	1.000	10.00	298.15

S17-14. Recalculate the reduction potential of the hydrogen electrode under the conditions stated below:

	pH	P_{H_2} (atm)	T (K)
(a)	1.000	1.00	298.15
(b)	−1.000	1.00	298.15
(c)	0.000	0.10	298.15
(d)	0.000	10.00	298.15

S17-15. A certain Zn/Cu cell exhibits a potential of 2.00 V at 25°C. **(a)** Calculate the reaction quotient. **(b)** Is the system in equilibrium? If not, in which direction will the reaction proceed—toward products or toward reactants?

S17-16. A solution containing Cr^{3+} ions is subjected to electrolysis for 100.0 minutes. What current must be applied to deposit 3.65 grams of chromium metal?

S17-17. How much time is needed to produce 10.0 grams of each metal by application of an electrolytic current of 2.00 A?

    (a) K from K^+ **(b)** Rb from Rb^+ **(c)** Mg from Mg^{2+} **(d)** Ti from Ti^{4+}

S17-18. How many moles of each metal will be obtained after 10.0 hours of electrolysis at 5.00 A?

    (a) Ag from Ag^+ **(b)** K from K^+ **(c)** Ni from Ni^{2+} **(d)** Cu from Cu^{2+}

S17-19. Which of the following species, under normal electrochemical conditions, is able to act simultaneously as an oxidizing agent and a reducing agent?

(a) Fe^{2+} (b) Cu   (c) Na^+

S17-20. (a) Write the following process as a sum of reduction and oxidation half-reactions:

$$2Cu^+(aq) \rightarrow Cu^{2+}(aq) + Cu(s)$$

(b) Which species is the oxidizing agent? Which is the reducing agent?

S17-21. The *disproportionation* reaction treated in the preceding exercise,

$$2Cu^+(aq) \rightarrow Cu^{2+}(aq) + Cu(s)$$

proceeds very rapidly in aqueous solution, usually coming to equilibrium in less than one second. (a) Use the formation data in Table C-16 to calculate the equilibrium constant at 25°C. (b) Which of two salts—CuCl or $CuCl_2$—would you expect to be insoluble in water? Explain your reasoning.

S17-22. Continue the analysis from the preceding exercise, and suppose (contrary to fact) that the reaction

$$2Cu^+(aq) \rightarrow Cu^{2+}(aq) + Cu(s)$$

proceeds very slowly, perhaps taking centuries to reach equilibrium. If so, would you still be prepared to comment on the relative solubilities of CuCl and $CuCl_2$?

S17-23. F_2 has a large, positive reduction potential. Li^+ has a large, negative reduction potential. At equilibrium, state whether the residual free energy (ΔG) for the reduction of fluorine,

$$F_2(g) + 2e^- \rightleftarrows 2F^-(aq)$$

is greater than, less than, or equal to the residual free energy for the reduction of lithium:

$$Li^+(aq) + e^- \rightleftarrows Li(s)$$

Be sure to distinguish ΔG from $\Delta G°$.

Chapter 18

Kinetics—The Course of Chemical Reactions

S18-1. Explain the meaning of the following terms, pointing out how each is different: *rate, rate law, rate constant*.

S18-2. Is it possible for the reaction

$$A + B \rightarrow C + D + E$$

to be described by one or more of the rate laws listed below? If so, which ones?

$$\text{Rate} = k[A]$$

$$\text{Rate} = k[A][B]$$

$$\text{Rate} = k[A]^2[B]$$

$$\text{Rate} = \frac{k_1[A]^{3/2}[C]}{[D]^{5/4} - k_2[E]} + k_3[A]$$

S18-3. **(a)** Assign correct units to the rate constant in the following expression:

$$\text{Rate} = k\frac{[A][B]}{[C]^2[D]^2}$$

(b) What is the overall order of reaction?

S18-4. Something new: If the kinetics of a reaction

$$A \rightarrow products$$

are *zeroth* order, the rate law is given by

$$Rate = -\frac{\Delta[A]}{\Delta t} = k$$

(a) What are the proper units for k? **(b)** If the initial concentration of A is $[A]_0$, convince yourself that the concentration at time t is given by

$$[A]_t = -kt + [A]_0$$

(c) Write an expression for the half-life. **(d)** At what time will $[A]_t$ fall to the value 0?

S18-5. Assume that the reaction

$$A \rightarrow products$$

follows zeroth-order kinetics, as described in the preceding exercise. Concentrations at $t = 10$ s and $t = 100$ s are found to be 1.00 M and 0.75 M, respectively. **(a)** Calculate k. **(b)** Calculate $[A]_0$. **(c)** Calculate $t_{1/2}$.

S18-6. Suppose that the initial rate of a hypothetical reaction

$$A + B + C \rightarrow D + E$$

is measured as follows:

TRIAL	[A]	[B]	[C]	INITIAL RATE ($M\,s^{-1}$)
1	0.10	0.10	0.10	0.484
2	0.20	0.10	0.10	0.483
3	0.10	0.20	0.10	0.967
4	0.10	0.10	0.20	1.936

All concentrations are in moles per liter (1 $M = 1$ mol L^{-1}). **(a)** Determine the kinetic order with respect to each reactant. **(b)** What is the overall rate law and order of reaction?

S18-7. Suppose that the reaction

$$2A \rightarrow products$$

obeys the following initial rate law:

$$Rate = k[A]^2$$

The initial concentration of A is 2.00 *M*, and the first half-life is 155.1 s. **(a)** Calculate the concentration at $t = 30.0$ s. **(b)** Is it possible for the reaction to be elementary? Can you tell?

S18-8. Say that the initial rate law for a reaction

$$A + B \rightarrow C$$

is second order in [A]:

$$Rate = k[A]^2$$

(a) Must this same rate law be maintained at all times? What additional variable or variables might become important as the process nears equilibrium? **(b)** Is it possible for the reaction to be elementary?

S18-9. Reaction 1 has an activation energy equal to 100 kJ mol^{-1}. For reaction 2, the activation energy is 10 kJ mol^{-1}. Further information, such as concerning collision frequency and efficiency, is not available. **(a)** Do you have sufficient data to predict which of the two reactions will have the larger rate constant at 25°C? If yes, do so. If no, specify what additional information is needed. **(b)** Do you have sufficient data to predict whether each reaction will be endothermic or exothermic? If yes, do so. If no, specify what else is needed. **(c)** Do you have sufficient data to predict whether the equilibrium constant for each reaction will be greater than 1, less than 1, or equal to 1? If yes, do so. If no, specify what more is needed.

S18-10. The rate constant $k(T)$ for a certain reaction obeys the Arrhenius law. Assume further that the pre-exponential factor is independent of temperature. **(a)** Does $k(T)$ increase, decrease, or remain the same as the temperature is increased? **(b)** Show that the relationship

$$\frac{1}{T_2} = \frac{1}{T_1} - \frac{R \ln p}{E_a}$$

holds when

$$\frac{k(T_2)}{k(T_1)} = p$$

S18-11. Use the results from Exercise S18-10 to answer the following questions.
(a) Suppose that $E_a = 10$ kJ mol^{-1}. At what temperature will the value of k be tripled relative to its value at $T_1 = 300$ K? **(b)** Suppose, instead, that $E_a = 100$ kJ mol^{-1}. At what temperature will the value of k be tripled relative to its value at $T_1 = 300$ K?

S18-12. The rate constant for a certain first-order reaction is 0.500 s^{-1} at 400 K.
(a) Calculate the half-life. **(b)** If the initial concentration is 0.750 M, what concentration remains after 3.25 s? **(c)** Do you expect the half-life to increase, decrease, or remain the same as the temperature is increased? **(d)** Do you expect the half-life to increase, decrease, or remain the same as the initial concentration is increased?

S18-13. Assume that the rate constant for some first-order reaction obeys the Arrhenius law. **(a)** Calculate the half-life at 500 K, given the following values for the pre-exponential factor (A) and the activation energy (E_a):

$$A = 1.00 \times 10^{12} \text{ s}^{-1} \qquad E_a = 50.0 \text{ kJ mol}^{-1}$$

(b) Calculate the half-life at 600 K.

S18-14. Carried out in the gas phase, the reaction

$$A + B \rightarrow C + D$$

was found to obey the rate law

$$\text{Rate} = k[A][B]$$

with $k = 10^3 \ M^{-1} \text{ s}^{-1}$. **(a)** A priori—given no other information—would you expect the rate law to have the same algebraic form and order if the reaction were carried out in solution? **(b)** Would you expect the numerical value of k to be the same in solution as it is in the gas phase? **(c)** Would you expect the mechanism to be the same in solution as it is in the gas phase?

S18-15. Suppose that the equilibrium constant for some elementary reaction

$$A \rightleftarrows B$$

is exceedingly large—as large as, say, $K = 10^{50}$. At equilibrium, is the rate of the reverse reaction effectively zero? Does the reverse reaction *never* occur?

S18-16. Here are the forward and reverse rate constants (k_+ and k_-) for two elementary reactions:

REACTION	k_+ (s^{-1})	k_- (s^{-1})
A $\rightleftarrows$ B	10	20
C $\rightleftarrows$ D	100	50

(a) Calculate the equilibrium constant for each reaction. **(b)** Calculate the overall rate of each reaction at equilibrium.

S18-17. Assume that reaction 1 is elementary,

$$1. \quad A \rightleftarrows B$$

whereas reaction 2 is not:

$$2. \quad 3A + 2B + C \rightleftarrows 4D + E + F$$

The equilibrium constants are $K_1 = 100$ and $K_2 = 0.01$, respectively, for the two processes. **(a)** At equilibrium, is the overall rate of reaction 2 greater than, less than, or equal to the rate of reaction 1? **(b)** Can you determine a numerical value for the rate constant of reaction 1 in the forward direction? If yes, do so. If no, explain why. **(c)** Can you determine a numerical value for the rate constant of reaction 2 in the forward direction? If yes, do so. If no, explain why.

S18-18. Listed below are the free energies of reaction and activation for two hypothetical transformations, each process beginning with the same set of reactants:

REACTION	$\Delta G°$ (kJ mol^{-1})	$\Delta G^{\ddagger}$ (kJ mol^{-1})
R $\rightleftarrows$ P_1	-100	100
R $\rightleftarrows$ P_2	-10	10

(a) Which reaction reaches equilibrium faster? **(b)** Which reaction yields the greater proportion of products at equilibrium?

S18-19. Answer the following questions, taking into account the thermodynamic data presented in Exercise S18-18. **(a)** Imagine that R, P_1, and P_2 come to a three-way equilibrium:

$$P_1 \rightleftarrows R \rightleftarrows P_2$$

Which product—P_1 or P_2—is found in the greater proportion at equilibrium? **(b)** Which product will predominate in the early stages of the three-way reaction? Is the process initially controlled by thermodynamics or kinetics?

S18-20. Again, consider the three species described in the preceding two exercises:

REACTION	ΔG° (kJ mol^{-1})	$\Delta G^\ddagger$ (kJ mol^{-1})
R $\rightleftarrows$ P$_1$	-100	100
R $\rightleftarrows$ P$_2$	-10	10

This time, however, imagine that the product P$_2$ (a gas) disappears from the system as soon as it is produced. **(a)** Is the process ever able to reach the system-wide equilibrium depicted below?

$$P_1 \rightleftarrows R \rightleftarrows P_2$$

(b) Do you expect the eventual amount of P$_2$ produced to be greater than, less than, or the same as the amount of P$_1$? Is the process controlled by thermodynamics or kinetics?

S18-21. An experimenter studying the hypothetical process

$$A + 2B + C \rightarrow D$$

finds that the average rate—as measured by the disappearance of A—is 0.001 $M\,s^{-1}$ during the first 1.0 s of reaction:

$$-\frac{\Delta[A]}{\Delta t} = 0.001 \ M\ s^{-1}$$

During this same interval, however, the product D fails to appear. The average initial rate—as measured by the *appearance* of D—is therefore zero:

$$\frac{\Delta[D]}{\Delta t} = 0$$

Suggest a reason for the inequality between these two rates.

S18-22. What's wrong with the picture below?

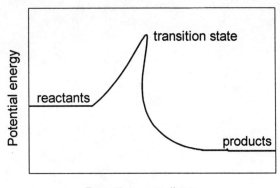

Chapter 19

Chemistry Coordinated—The Transition Metals and Their Complexes

S19-1. Analyze the coordination compound $K_2[Ni(en)Cl_4]$ as follows: **(a)** Identify the coordinated metal and its oxidation state. **(b)** Identify the ligands and deduce the coordination number. **(c)** Describe the geometric structure.

S19-2. Analyze the coordination compound $[Co(en)_3]Cl_3$ as follows: **(a)** Identify the coordinated metal and its oxidation state. **(b)** Identify the ligands and deduce the coordination number. **(c)** Describe the geometric structure.

S19-3. **(a)** Which one of these two complex ions—$[Co(NH_3)_6]^{3+}$ or $[CoF_6]^{3-}$—is likely to be paramagnetic? Which one is likely to be diamagnetic? **(b)** For each structure, use crystal field theory to predict the electron configuration.

S19-4. The complex ion $[AuCl_4]^-$ is square planar. **(a)** What is the oxidation state of the gold ion? **(b)** How many d electrons are present? **(c)** Is the complex ion paramagnetic or diamagnetic?

S19-5. **(a)** Suppose that a material absorbs light broadly at wavelengths between 360 nm and 780 nm. What color—red, green, violet, gray-black, or no color at all—do you expect the material to have? **(b)** What color do you expect for a material that absorbs light primarily at a wavelength of 300 nm? **(c)** What color do you expect for a material that absorbs light primarily at a wavelength of 850 nm?

S19-6. An aqueous solution of $[Co(H_2O)_6]^{2+}$ absorbs light at a frequency of 5.77×10^{14} Hz. **(a)** Calculate the wavelength of absorption. In which portion of the electromagnetic spectrum—infrared, visible, or ultraviolet—does the radiation fall? **(b)** Calculate the crystal field splitting parameter Δ, expressing your answer in kJ mol^{-1}.

S19-7. Which one of these two complexes—$[Co(NH_3)_6]^{3+}$ or $[Co(NH_3)_5H_2O]^{3+}$—absorbs light at the longer wavelength?

S19-8. Which one of these two complexes—$[Cd(NH_3)_6]^{2+}$ or $[Co(NH_3)_6]^{2+}$—is colorless?

S19-9. Is $[Zn(H_2O)_4]^{2+}$ colored or colorless?

S19-10. Solutions containing the complex ion $[Co(NH_3)_6]^{3+}$ are yellow. In which range of wavelengths (see below) do you expect the structure to absorb light?

$$400 \text{ nm} - 450 \text{ nm} \qquad 500 \text{ nm} - 550 \text{ nm} \qquad 600 \text{ nm} - 650 \text{ nm}$$

S19-11. How many d electrons are present in the Sc^{3+} ion? Do you expect an octahedral complex built around Sc^{3+} to be especially stable or unstable?

S19-12. **(a)** How many d electrons are present in Ti^{4+} and Mn^{7+}? **(b)** Which one of the two ions do you expect to be more stable? **(c)** Which one of these two species—Mn^{7+} or MnO_4^-—do you expect to be more stable?

S19-13. The ions Zn^{2+} and Ag^+ are especially stable. Explain why.

S19-14. Which one of these two complexes—$[Ag(NH_3)_2]^+$ or $[Ag(CN)_2]^-$—is thermodynamically more stable? Relevant data may be found in Appendix C of *Principles of Chemistry*.

S19-15. Calculate $\Delta G°$ for the formation of $[PbCl_4]^{2-}$:

$$Pb^{2+}(aq) + 4Cl^-(aq) \rightleftarrows [PbCl_4]^{2-}(aq)$$

Relevant data may be found in Appendix C.

S19-16. Calculate the equilibrium concentrations of Ag^+ and $S_2O_3^{2-}$ that result when 0.500 mol $[Ag(S_2O_3)_2]^{3-}$ is dissolved in 0.250 L H_2O. See the reaction below:

$$Ag^+(aq) + 2S_2O_3^{2-}(aq) \rightleftarrows [Ag(S_2O_3)_2]^{3-}(aq) \qquad K_f = 2.9 \times 10^{13}$$

S19-17. Use the following questions to draw an analogy between the heterogeneous equilibrium of an ionic solute AB_n,

$$AB_n(s) \rightleftarrows A^{n+}(aq) + nB^-(aq)$$

and the equilibrium between a coordination complex ML_n and its ligands:

$$ML_n(aq) \rightleftarrows M(aq) + nL(aq)$$

(a) Write an expression for the equilibrium constant in each system. (b) Which is the ordered subsystem and which is the disordered subsystem in each transformation? (c) Does your general picture apply equally well to the dissolution–precipitation equilibrium of a molecular solute X? See below:

$$X(s) \rightleftarrows X(aq)$$

S19-18. The formation constant for $[Zn(OH)_4]^{2-}$ is 2.8×10^{15} in aqueous solution. The formation constant for $[Zn(NH_3)_4]^{2+}$ is 2.9×10^9. (a) Can you predict which of the two complex ions will form faster? (b) Suppose that ammonia molecules and hydroxide ions compete simultaneously for the same zinc ions. Can you predict which one of the two complex ions will predominate at equilibrium?

S19-19. (a) Explain the difference between a thermodynamically *stable* complex and a kinetically *inert* complex. (b) Which quantity—the free energy of formation or the free energy of activation—determines stability and instability? (c) Which quantity determines inertness and lability?

S19-20. Octahedral complexes built around Cr^{3+} generally are both thermodynamically stable and kinetically inert compared with those built around Cr^{2+}. Explain why.

S19-21. Nearly all complexes of Cr^{3+} are octahedral. Explain why tetrahedral and square planar complexes are unfavorable.

S19-22. Do you expect Cr^{2+} to be a reducing agent or an oxidizing agent?

S19-23. Suppose that a certain complex is soluble in both water and ethanol. (a) Will the formation constant necessarily be the same? (b) Will the lability of the complex necessarily be the same? Explain why or why not.

Chapter 20

Spectroscopy and Analysis

S20-1. Match the items on the left with the items on the right:

gamma rays   molecular rotation
infrared radiation   nuclear spin
microwave radiation   molecular vibration
ultraviolet/visible radiation   core electrons
X rays   protons and neutrons
radiofrequency radiation   valence electrons

S20-2. Describe (or sketch) the ^{1}H NMR spectrum of each structure proposed below, paying attention to the number of lines and their relative intensities:

(a) HCCH

(b) HCCBr

(c) $CH_3CBr_2CBr_2CH_3$

(d) $CH_3CBr_2CBr_2CCl_2H$

Rewrite each of the formulas to make every connection and every bond explicit.

S20-3. Similar—describe or sketch the ^{1}H NMR spectrum of each proposed structure:

(a) CH_2BrCBr_2H

(b) CH_2Br_2

(c) CH_2CBr_2

(d) *trans*-CHBrCHCl

Rewrite the formulas to make every connection and every bond explicit.

S20-4. Carbon-13, which makes up only 1.1% of naturally occurring carbon, has a nuclear spin quantum number of $\frac{1}{2}$. The abundant ^{12}C isotope, by contrast, has a nuclear spin quantum number of 0. **(a)** Sketch the ^{13}C NMR spectrum of $^{12}CCl_3{}^{13}CH_2{}^{12}CCl_3$, taking into account the local field produced by the hydrogen nuclei. **(b)** How would the ^{13}C spectrum change if the hydrogen nuclei did not contribute a local field?

S20-5. The mass spectrum of chromium contains four signals:

RELATIVE MASS	INTENSITY
49.946046	0.05186
51.940509	1.00000
52.940651	0.11339
53.938882	0.02823

(a) To which isotope does each signal correspond? (Note that the mass of an atomic nucleus is slightly less than the mass of its protons and neutrons taken separately. See Chapter 21.) **(b)** Calculate the average molar mass of chromium, using only the data supplied here. **(c)** Suppose that a naturally occurring sample of chromium contains 20.00 g. How many grams of ^{53}Cr are present in the sample?

S20-6. Sketch the mass spectrum of lead, consulting Table C-11 as needed.

S20-7. The mass spectrum of $C_{18}H_{38}$ contains many signals, including a peak corresponding to a mass/charge ratio of 254. Do you expect this signal to be the most intense in the spectrum? Why or why not?

S20-8. Transitions between rotational energy levels usually occur when an oscillating electric field interacts with a permanent electric dipole moment on a structure. Which of the following species are "microwave-active" (able to produce a rotational spectrum)?

    (a) H_2O　　**(b)** CH_4　　**(c)** HCl　　**(d)** CO_2

    (e) NH_3　　**(f)** Ar　　**(g)** CH_3Cl

S20-9. A nonpolar molecule, despite the absence of a permanent electric dipole moment, sometimes can produce a rotational spectrum. To do so, the structure must possess a permanent *magnetic* dipole moment able to interact with a magnetic field. Which one of these diatomic molecules—H_2, Li_2, or O_2 —therefore might be microwave-active in its ground state? If necessary, review the discussion of diatomic molecular orbitals in Chapter 7.

S20-10. Transitions between vibrational levels usually occur when a dipole moment *changes* while it interacts with an electric field. Which of the following modes of vibration are "infrared-active" and thus will appear in a vibrational spectrum?

(a) stretching of OH in H_2O

(b) bending of HOH in H_2O

(c) stretching of H_2

(d) stretching of HCl

(e) symmetric stretching of OCO in CO_2

(f) bending of OCO in CO_2

Hint: During the course of a "symmetric stretch," the lengths of the two CO bonds in CO_2 increase and decrease in tandem—equally on either side of the molecule. The OCO bond angle remains unchanged during a symmetric stretching vibration, but the angle varies periodically during a bending vibration.

S20-11. When an excited helium atom undergoes a transition from the $1s^12p^1$ state down to the ground state ($1s^2$), it emits electromagnetic radiation at a wavelength of 58.43 nm—the so-called He(I) line. **(a)** In what portion of the electromagnetic spectrum does the He(I) line fall? **(b)** What is the corresponding frequency? **(c)** What is the corresponding photon energy?

S20-12. Here is an example of *ultraviolet photoelectron spectroscopy*: **(a)** Write the valence electron configuration of N_2, referring back to Chapter 7 as needed. Is the highest occupied molecular orbital (abbreviated HOMO) of σ symmetry or π symmetry? **(b)** The ionization energy of an electron in the HOMO of N_2 is 1504 kJ mol^{-1}. If this electron is ionized by He(I) radiation (see the preceding exercise), with what kinetic energy will it emerge? **(c)** What will be its speed?

S20-13. When excited by a wavelength of 58.43 nm, as above, N_2 also ejects an electron from an orbital lying just below the HOMO. **(a)** The speed of this additional photoelectron is 1.262×10^6 m s^{-1}. What is the corresponding kinetic energy? **(b)** Calculate the ionization energy in kJ mol^{-1}.

S20-14. Do you expect ultraviolet photoelectron spectroscopy, as illustrated in the preceding two examples, to be generally useful for studying electrons in the core of an atom or molecule?

S20-15. The $1s$ ionization energies for three second-row atoms are listed below:

(a) Li 4.82×10^3 kJ mol^{-1}

(b) B 1.83×10^4 kJ mol^{-1}

(c) F 6.66×10^4 kJ mol^{-1}

Calculate the wavelength needed to eject a core electron from each atom. In what portion of the electromagnetic spectrum does each of the energies lie?

Chapter 21

Worlds Within Worlds—The Nucleus and Beyond

S21-1. Relativistic effects usually scale in proportion to a factor

$$\gamma = \frac{1}{\sqrt{1 - \dfrac{v^2}{c^2}}}$$

in which c denotes the speed of light and v denotes the velocity of a uniformly moving reference frame. **(a)** Evaluate γ for a particle at rest in a given inertial frame. **(b)** Recalculate γ for the same particle moving with velocity $v = 3 \times 10^6$ m s^{-1}, one-hundredth the speed of light. **(c)** Repeat for $v = 0.1c$, $0.5c$, $0.9c$, $0.99c$, and $0.999c$.

S21-2. The relativistic kinetic energy of a particle with uniform velocity v is given by the following formula:

$$E = \frac{mc^2}{\sqrt{1 - \dfrac{v^2}{c^2}}} \approx mc^2 + \frac{1}{2}mv^2 + \cdots$$

The *rest mass* of the particle, m, is constant in all inertial reference frames. **(a)** Calculate E for a free electron moving at $0.001c$. **(b)** Calculate the Einsteinian *rest energy* of the particle, mc^2. **(c)** Calculate the nonrelativistic (Newtonian) kinetic energy, $\frac{1}{2}mv^2$. **(d)** How well is the relativistic energy at this speed approximated by the sum of the rest energy and the nonrelativistic kinetic energy?

S21-3. Repeat the preceding exercise for a free electron traveling now at $0.5c$. **(a)** Calculate the relativistic energy. **(b)** Calculate the rest energy. **(c)** Calculate the nonrelativistic kinetic energy. **(d)** Compare the relativistic energy at this speed with the sum of the rest energy and kinetic energy.

S21-4. Once more, this time for a free electron traveling at $0.999c$. **(a)** Calculate the relativistic energy. **(b)** Calculate the rest energy. **(c)** Calculate the nonrelativistic kinetic energy. **(d)** Compare the relativistic energy with the sum of the rest energy and nonrelativistic kinetic energy.

S21-5. **(a)** For which electron—a $1s$ electron in carbon or a $1s$ electron in gold—would relativistic effects be more likely to appear? **(b)** Again: For which electron—a $1s$ electron in gold or a $7s$ electron in plutonium—would relativistic effects be more likely to appear?

S21-6. Suppose that a particle has a nonzero rest mass. What would be its relativistic energy if it moved at the speed of light? Can it do so?

S21-7. The photon has zero rest mass and moves at the speed of light: $v = c$. Yet despite its zero *rest* mass, the photon is still subject to gravitational interactions. A beam of light, for example, will bend in the presence of a strong gravitational field, such as might be produced by a dense star. Are you surprised, given that we customarily think of gravity as a "force" arising from the interaction of two *masses*? Use a very rough, very brief relativistic argument to make plausible the idea that a "massless" particle can be affected by gravity.

S21-8. **(a)** Calculate Δm and ΔE for the nuclear fusion reaction given below:

$$\mathrm{^{1}_{1}H} + \mathrm{^{1}_{1}H} \rightarrow \mathrm{^{2}_{1}H} + \mathrm{^{0}_{1}e} + \nu$$

Use the value 2.014102 u for the atomic mass of deuterium. **(b)** Is the process endothermic or exothermic? Is the transformation favorable from a thermodynamic standpoint?

S21-9. Continue with the fusion reaction described in the preceding exercise, with an eye now toward kinetic rather than thermodynamic benchmarks. **(a)** Start with the formula for the Coulomb energy between charges q_1 and q_2 interacting over a distance r:

$$E = \frac{1}{4\pi\varepsilon_0}\frac{q_1 q_2}{r}$$

Use this equation to calculate the potential energy of two hydrogen nuclei separated by 2×10^{-15} m, the distance at which the strong force begins to take hold. Values of the physical constants are tabulated in Appendix C. **(b)** Think of this repulsive Coulomb potential as a kinetic activation barrier. At approximately what temperature would two hydrogen nuclei have sufficient thermal energy to overcome the barrier? (Remember, from Chapter 10, that $\frac{3}{2}k_\mathrm{B}T$ is the average translational kinetic energy of a particle.) **(c)** The answer you have just obtained may well be an overestimate, but ask nonetheless: Is the reaction likely to occur under ordinary terrestrial conditions? **(d)** For comparison, calculate the temperature at which the average translational kinetic energy of a particle would be 50 kJ mol^{-1}—a typical activation barrier for a chemical reaction.

S21-10. The element neon, a noble gas, is unreactive chemically. Does its nobility also extend to nuclear reactions? Explain the difference between a chemical reaction and a nuclear reaction.

S21-11. Different isotopes of the same element exhibit nearly the same chemical behavior. Do two isotopes—say uranium-235 and uranium-238—also behave the same way in nuclear reactions? Why or why not?

S21-12. **(a)** Is a typical nuclear reaction affected by the immediate chemical environment around the nucleus? Will a particular nucleus react differently when incorporated into different molecules? **(b)** Can you think of a possible exception to the answer you just gave?

S21-13. **(a)** Calculate Δm and ΔE for the nuclear transformation given below:

$$\prescript{1}{0}{n} + \prescript{235}{92}{U} \rightarrow \prescript{142}{56}{Ba} + \prescript{91}{36}{Kr} + 3\,\prescript{1}{0}{n}$$

The atomic masses of barium-142 and krypton-91 are 141.916360 u and 90.923380 u, respectively. Other relevant data will be found in Appendix C. **(b)** What kind of reaction does the equation describe? **(c)** Is the process endothermic or exothermic? Is it favorable in a thermodynamic sense? **(d)** Is the kinetic activation barrier higher or lower here compared with a fusion reaction? Explain.

S21-14. **(a)** Calculate the energy produced by the annihilation of 1.00×10^{-20} g of electrons and 1.00×10^{-20} g of positrons. **(b)** Calculate the annihilation energy for 1.00×10^{-20} g of protons and 1.00×10^{-20} g of antiprotons.

S21-15. Science fiction or fact: If antimatter could be produced economically and in sufficient quantity, might the process of annihilation offer a practical source of energy? **(a)** Calculate the energy needed to illuminate 10 million 60-watt bulbs for an entire year. Note that the watt (W) is a measure of *power*, energy radiated per unit time: $1 \text{ W} = 1 \text{ J s}^{-1}$. **(b)** What total mass of positrons and electrons (expressed in kilograms) would be sufficient to keep the lights burning?

S21-16. An example of the tremendous energy involved in a typical nuclear fusion process: **(a)** The atomic mass of deuterium is 2.014102 u. Calculate the energy released during the particular reaction shown below, expressing the result in units of kJ mol^{-1}:

$$\prescript{2}{1}{H} + \prescript{3}{1}{H} \rightarrow \prescript{4}{2}{He} + \prescript{1}{0}{n}$$

(b) Suppose that water fills a cubical tank 25.0 meters on a side. If all the energy produced in the making of 1.000 mol ^{4}He were absorbed by the water at 25°C, to what temperature would the liquid rise? Assume that H_2O has a density of 1.00 g cm^{-3} and a heat capacity of 75.3 J mol^{-1} K^{-1} at constant pressure. **(c)** By comparison, what would be the final temperature if (as in a typical chemical reaction) the water absorbed 100 kJ?

S21-17. Don't forget! Strictly speaking, mass is *not* conserved in a chemical reaction. Wherever there is a change in energy, there is a change in mass—but the change in mass brought about by a chemical reaction is so slight as to be practically negligible. **(a)** Calculate ΔE and Δm for a typical chemical transformation, the combustion of one molecule of methane:

$$CH_4(g) + 2O_2(g) \rightarrow CO_2(g) + 2H_2O(g)$$

Hint: Assume that $\Delta E \approx \Delta H$, and then work backward from the equation $\Delta E = (\Delta m)c^2$ to determine Δm. Relevant data may be found in Appendix C. **(b)** Express your value of Δm as a percent change relative to the mass of reactants.

S21-18. Consider the chemical reaction

$$N_2O_4(g) \rightarrow 2NO_2(g)$$

How much mass does the system gain or lose as one molecule of N_2O_4 is converted into two molecules of NO_2? Use the same method as in the preceding example.

S21-19. Is it possible for berkelium-244 to decay into lead-207 through a series of alpha and beta emissions?

S21-20. The difference in energy between the nuclear ground states of $^{27}_{12}Mg$ and $^{27}_{13}Al$ is 4.18×10^{-13} J. **(a)** By emission of a β^- particle, a nucleus of magnesium-27 decays first to an excited state of aluminum-27. The excited nucleus then drops to the ground state by emitting a γ photon with a frequency of 2.44×10^{20} Hz. Calculate the energy of the β^- particle and the wavelength of the γ photon, assuming (for simplicity) that no energy is lost through other channels. **(b)** In an alternative route of decay, magnesium-27 emits a β^- particle having an energy of 2.80×10^{-13} J. Estimate the energy and wavelength of the γ photon subsequently emitted, making the same simplifying assumption as before. **(c)** Comment on the wavelengths and energies of the γ radiation.

S21-21. Classify each particle as a lepton, baryon, or meson:

(a) $u\tilde{d}$ **(b)** $\tilde{\nu}$ **(c)** $\tilde{u}\tilde{u}\tilde{d}$ **(d)** ^1_0n **(e)** ^0_1e

S21-22. Identify the particles in each of the following pairs:

(a) udd, $\tilde{u}\tilde{u}\tilde{d}$ **(b)** $^0_{-1}e$, ^1_1p **(c)** ν, $\tilde{\nu}$ **(d)** udd, $\tilde{u}\tilde{d}\tilde{d}$

Which pairs can undergo annihilation?